W9-BPL-751

UNSCIENTIFIC AMERICA

UNSCIENTIFIC AMERICA

How Scientific Illiteracy Threatens Our Future

CHRIS MOONEY
AND
SHERIL KIRSHENBAUM

BASIC
BOOKS

A Member of the Perseus Books Group
New York

Published by Basic Books,
A Member of the Perseus Books Group

· Books published by Basic Books are available at special discounts for bulk purchases in the United States by corporations, institutions, and other organizations. For more information, please contact the Special Markets Department at the Perseus Books Group, 2300 Chestnut Street, Suite 200, Philadelphia, PA 19103, or call (800) 810-4145, ext. 5000, or e-mail special.markets@perseusbooks.com.

Designed by Brent Wilcox

Library of Congress Cataloging-in-Publication Data
Mooney, Chris.
    Unscientific America : how scientific illiteracy threatens our future / Chris Mooney and Sheril Kirshenbaum.
        p.   cm.
    Includes bibliographical references and index.
    ISBN 978-0-465-01305-0 (alk. paper)
    1. Science—Study and teaching—United States.   2. Communication in science—United States.   3. Science in popular culture—United States.
I. Kirshenbaum, Sheril.   II. Title.
    Q149.U5M66 2009
    509.73—dc22
                                                                    2009015482

10 9 8 7 6 5 4 3 2 1

Of course we're intimidated by science! Science holds
itself above everybody else—above God, evidently.
You guys have been kicking ass since the Enlightenment.

—STEPHEN COLBERT

Instead of being derided as geeks or nerds, scientists
should be seen as courageous realists and the last great
heroic explorers of the unknown. They should get more
money, more publicity, better clothes, more sex and free
rehab when the fame goes to their heads.

—MATTHEW CHAPMAN,
screenwriter and cofounder, ScienceDebate2008

CONTENTS

PART III   THE FUTURE IN OUR BONES

FROM A SCIENTIST AND A WRITER

If we're successful, this book will seamlessly merge form and content. For it is the collaborative work of a writer and a scientist, and it argues that we need many more such "two cultures" partnerships if we're to forge the connections between American science and American society that will guide us through the twenty-first century.

Chris is a journalist who learned to value science's humbling lessons and penetrating way of thinking at a young age. His biologist grandfather, Gerald Cole, had a powerful influence: "Paw" liked to refer to Charles Darwin as "Chuck" and pretend he was sitting right there at the dinner table. Chris's first book, *The Republican War on Science*, took up the family tradition and helped feed a growing awareness of the ways in which science has been abused in the political realm, thereby jeopardizing our ability to address pressing issues such as global warming. But over time, Chris came to see that the problematic status of science in our society sprang from causes far more diverse than the most immediate one (conservative ideologues attacking well-established knowledge) and that the solution required far more than throwing George W. Bush out of the White House. In particular, he began to write and lecture about the need for scientists to communicate their knowledge in ways that non-scientists can relate to and understand.

Sheril took a very different trajectory, yet converged on a similar place. Currently an associate at Duke University, she holds two master of science degrees in marine biology and marine policy from the University of Maine, where she studied the population dynamics and life

history of *Cucumaria frondosa*—the ever-charismatic sea cucumber—
and worked with the fishing community to preserve and manage the
species. Sheril continues to publish in scientific journals, but instead of
pursuing a Ph.D. she accepted a position on Capitol Hill working with
Senator Bill Nelson (D-FL) on energy, climate, and ocean policy. Far
from the ivory tower, Sheril soon saw how difficult it can be to integrate science into the public policy process and how often scientists fail
to connect with top decision makers. A stint working in pop radio as a
"Top 40" DJ, meanwhile, showed her how jocks engage the public
using basic social-marketing techniques and convinced her that the
world of science might get a shot in the arm from employing similar
strategies on occasion.

Both of our careers, then, have drawn upon the creative energy generated at the intersection between science and other disciplines or approaches. The central inspiration for this book was precisely such a
culture-crossing case study: ScienceDebate2008, an initiative in which
we joined up with two Hollywood screenwriters, a physicist, a lawyer,
and a philosopher to try something unheard of—mobilizing the American scientific community to demand that politicians address crucial
matters of science policy on the campaign trail. Within months we had
dozens of Nobel laureates, scores of scientific luminaries, over 100 university presidents, a wide range of scientific institutions and societies,
and 38,000 individuals supporting us, an unprecedented response from
the traditionally staid science world. But although the initiative had
many positive repercussions, politicians from both parties largely managed to ignore us during the campaign. So did the mass media. It was
quite a wake-up call, and demonstrates just how far we—and they—still
have to go.

Yet through countless discussions about the place of science in our
politics and our culture, we've developed the conviction that a better
future is possible and that we can build on undertakings like Sci-
enceDebate2008 to help ensure it. If we're to meet the science-based
challenges that will dominate this century, we have no other choice.

The good news is that President Barack Obama's administration, with a Nobel laureate as secretary of energy, a restored White House science adviser, and many other distinguished researchers in positions of major influence, represents a dramatic step forward for science and its role in public life. The "reality-based community" has been reinstated in Washington; after the Bush administration and its "war on science," it feels like a sunrise. Yet we can't expect the long-standing gap between scientists and the broader American public to disappear overnight, meaning this is no time for satisfaction or complacency. If the metaphorical "war" on science is over, now's the time for the long and difficult process of "nation building"—for laying sounder foundations to ensure it doesn't come raging back.

And not a moment too soon: Even as science is crucial to the fate of twenty-first-century America, it's under assault from new forces that not even the science-friendly Obama administration can fully address, because they're as much cultural and economic as directly political in nature. This book details what we consider the main challenges, centering on the immense difficulty of bringing useful and accurate information about science to our political and cultural leaders and to the broader American public, a long-standing communication problem that only appears to be growing more grave and urgent. Yet we find hope in perhaps the most unexpected of places: The army of young researchers on campuses across the country who do not want to be *just* scientists, but instead nourish a powerful desire to reach out to the society in which they live, and to which they owe so much.

Our deepest aspiration is that this book will push these young scientists, and those who share their enthusiasm and sense of mission, along that path. They are the future, and we need their help to break down the walls that have for too long separated the "experts" from everybody else. If we can combine the restoration of science in Washington with a renewed effort, partly grassroots in nature, to reconnect it with our broader society, perhaps we can finally create a stronger rapport between American science and mainstream American culture.

Right now the public needs that very badly, but so too do the scientists.

Writing a book is a long and yet at times frantic process, and we couldn't have gotten through it alone. For helpful readings, feedback, and copious useful information and advice, we'd like to thank Glenn Branch, D. Graham Burnett, Darlene Cavalier, Matthew Chapman, Jonathan M. Gitlin, Kei Koizumi, Sriram Kosuri, David Lowry, Molly McGrath, Sally Mooney, Shawn Lawrence Otto, Robert Pennock, Stuart Pimm, Phil Plait, Andrew Plemmons Pratt, Eric Roston, Reece Rushing, Paul Starr, and Al Teich. For putting us on a work schedule, we're indebted to Michelle Foncannon; and for helping us see how to unlock our ideas, to Sydelle Kramer and Bill Frucht, and to Lara Heimert, whose judicious edits were a revelation and who made us realize that we could say far more with vastly fewer words.

Chris also wishes to thank the Center for Inquiry West, in Hollywood, for allowing him to use its work space, and the Center for American Progress's Science Progress Web site (http://www.scienceprogress.org) for the opportunity to test-drive many of the ideas that eventually fused into this book. And he wants to specially thank Matthew Nisbet, who opened his eyes to a revealing body of research on the communication of science that has informed and enriched this project. A series of nationwide lectures they gave together in 2007 and 2008 served as an occasion for thinking through some of the arguments advanced here, and although they do not always agree—especially about ScienceDebate 2008—Chris is indebted to Nisbet for many enlightening conversations and dialogues, as well as for his comments on an early draft of this book. Additional thanks go to filmmaker Randy Olson, whose films about science communication (*Flock of Dodos* and *Sizzle!*) have been deeply thought-provoking, who read and commented on our Hollywood chapter, and whose forthcoming book, *Don't Be Such a Scientist: Talking Substance in an Age of Style*, resonates with our own project. Finally, on a personal note, Chris wants to thank his fiancée, Molly

McGrath, for her faith, support, and refusal to let him work and be serious all the time; and his Boston terrier, Sydney, for understanding that Daddy couldn't go on as many walks as usual when the book deadlines came up.

Sheril would like to thank the Pimm group and members of the Duke community for work space and stimulating conversations that enriched the pages that follow. She wishes to thank David Lowry for constant encouragement, inspiration, and excellent cooking throughout composition of this book, Vanessa Woods for endless advice, Rebecca Katof for unconditional support, Megan Dawson for holding the band together in her absence, and Nicolas Devos for his ever-optimistic outlook. Thanks finally to Mom, Dad, Seth and Rose Kirshenbaum, Jen Kiok, Sea Grant Fellows past and present, and everyone who has motivated her along the journey.

Last but hardly least: We want to dedicate this book to the core ScienceDebate2008 crew—Erik Beeler, Darlene Cavalier, Matthew Chapman, Austin Dacey, Lawrence Krauss, and Shawn Lawrence Otto—who constantly inspire us and who prove, to a very high degree of certainty, that any initiative can succeed if only it has the right people behind it. Granted, a little funding also helps, and we're pleased to announce that we'll be devoting a fixed percentage of royalties from sales of this book to ScienceDebate. Here's to 2012!

Chris Mooney and Sheril Kirshenbaum,
May 2009

CHAPTER 1

Why Pluto Matters

"Viva Pluto!"

"Stop Planetary Discrimination!"

"Pluto Was Framed!"

"Dear Earth: You Suck. Love, Pluto."

"Pluto is still a planet. Bitches."

THUS READ A SMALL SAMPLING OF DEFIANT T-SHIRT AND BUMPER STICKER slogans after the general assembly of the International Astronomical Union (IAU), meeting in Prague in late 2006, voted to excommunicate the ninth planet from the solar system. The union's action abruptly stripped Pluto of a status as much cultural, historical, and even mythological as scientific.

In the astronomers' defense, it had become increasingly difficult to justify calling Pluto a planet without doing the same for several other more recently discovered heavenly objects, one of which, the distant freezing rock now known as Eris (formerly "Xena"), turns out to be larger. But that didn't mean the experts had to fire Pluto from its previous place in the firmament. In defining the word *planet,* they were arguably engaged not so much in science as in semantic exercise. Instead of ruling Pluto out, they could just as easily have ruled a few new planets in, as a group of scientists, historians, and journalists had in fact proposed. But the IAU rejected that compromise for a variety of technical reasons: Pluto is much smaller than the other eight planets; it orbits the sun in a far more elliptical manner; its gravitational pull is not strong enough to have "cleared the neighborhood around its orbit" of other significant objects and debris; and so forth.

People were aghast. Not only did they recoil at having to unlearn a childhood science lesson, perhaps the chief thing they remembered about astronomy. On some fundamental level their sense of fair play had been violated, their love of the underdog provoked. Why suddenly kick Pluto out of the planet fraternity after letting it stay in for nearly a century, ever since its 1930 discovery? "No do-overs," wrote one cartoonist.

Soon, newly launched Web sites began encouraging people to vote on Pluto's status and override the experts. A Facebook group entitled "When I was your age, Pluto was a planet" drew in 1.5 million members. New Mexico, the state where Pluto's discoverer, Clyde Tombaugh, had built an astronomy program, took particular offense. Its House of Representatives voted unanimously to preserve Pluto's planethood and named March 13, 2007, "Pluto Planet Day." Surveying it all, the American Dialect Society selected "plutoed" as its 2006 word of the year—as in, "You plutoed me." The society offered this definition: "to demote or devalue someone or something, as happened to the former planet Pluto when the General Assembly of the International Astronomical Union decided Pluto no longer met its definition of a planet."

Even many scientists were upset. "I'm embarrassed for astronomy," remarked Alan Stern, the chief scientist on NASA's New Horizons mission to Pluto and beyond. Stern questioned the legitimacy of the Pluto demotion process: "Less than 5 percent of the world's astronomers voted," he charged. Other experts also dissented, even as some wags dubbed the IAU the "Irrelevant Astronomical Union." Comedians had a field day. Science had opted to "cut and run" on Pluto, quipped Bill Maher. The onetime planet had been forced to join its "own kind" in the outer solar system, "separate but equal," added Stephen Colbert. There were countless other jokes, many of which made the scientific community, supposedly calm and hyperrational, sound more than a little capricious in this instance.

Ultimately, the Pluto decision pleased almost no one; it may even be redebated at the next IAU meeting, slated for August 2009 in Rio de Janeiro. But if that's the case, how could this planetary crack-up happen in the first place? Didn't the scientists involved foresee such a public outcry? Did they simply not care? Was the Pluto decision really scientifically necessary?

Such questions implicate far more than our current conception of the solar system, or which planets babies will see in the mobiles overhanging their cribs. The furor over Pluto is just one particularly colorful example of the rift today between the world of science and the rest of society. This divide is especially pronounced in the United States, which is simultaneously the world's scientific leader—at least for the moment—and home to an overarching culture that often barely seems to know or care. (Unless scientists mess with Pluto, that is.)

It's a stunning contradiction, when you think about it. The United States features a massive infrastructure for science, supported by well over $100 billion annually in federal funding and sporting a vast network of government laboratories and agencies, the finest universities in the world, and innovative corporations that conduct extensive research. Thanks to such investments, Americans built the bomb, reached the moon, decoded the genome, and created the Internet. And yet today this country is also home to a populace that, to an alarming extent, ignores scientific advances or outright rejects scientific principles. A distressingly large number of Americans refuse to accept either the fact or the theory of evolution, the scientifically undisputed explanation of the origin of our species and the diversity of life on Earth. An influential sector of the populace is in dangerous retreat from the standard use of childhood vaccinations, one of medicine's greatest and most successful advances: By the end of the twentieth century, they were responsible for saving a million lives per year. The nation itself has become politically divided over the nature of reality, such that college-educated Democrats are now more than twice as likely as college-educated Republicans to believe that global warming is real and is caused by human activities. Meanwhile, the United States stands on the verge of falling behind other nations such as India and China in the race to lead the world in scientific endeavor in the twenty-first century.

If we allow that final lapse to occur, surely part of the reason will be that most of our citizens have had only fleeting encounters with a world of science that can appear baffling, intimidating, and even downright unfriendly. Just 18 percent of Americans know a scientist personally, according to survey data, and even fewer can name the government's top scientific agencies: the National Institutes of Health (NIH) and the National Science Foundation (NSF). When polled in late 2007 and asked to name scientific role models, 44 percent of the respondents didn't have a clue. They simply couldn't give an answer. And among those polled who did respond, the top selections were Bill Gates, Al Gore, and Albert Einstein, people who are either not scientists or not alive.

It's no wonder, then, that even as our scientists get up each morning and resume the task of remaking the world, the American public all too rarely follows along. This alienation leads to recurrent flare-ups like the Pluto episode, in which people suddenly catch wind of what scientists have been doing and react with anger, alarm, or worse.

The snubbing of Pluto won't have dire consequences back here on Earth, but other consequences of the science-society divide may prove far more damaging. We live in a time of climatic change and energy crisis, of widespread ecological despoilment and controversial biomedical research. We have great cause to fear global pandemics, nuclear proliferation, and attacks by tech-savvy terrorists. We stand on the verge of pathbreaking discoveries in genetics and neuroscience (to name just a few fields) that could redefine who we are and even upend our society. This is a time when science is pivotal to our political lives, our prosperity, and even our lifestyles and habits. And yet again and again, we encounter disturbing disconnects between the state of scientific understanding and the way we live our lives, set our policies, define our identities, and inform and entertain ourselves.

The problem isn't merely the dramatic cultural gap between scientists and the broader American public. It's the way this disconnect becomes self-reinforcing, even magnified, when it resurfaces in key sectors of society that powerfully shape the way we think, and where science ought to have far more influence than it actually does—in politics, the news media, the entertainment industry, and the religious community.

In the political arena from 2001 through 2008, the United States was governed by an administration widely denounced for a disdain of science unprecedented in modern American history. Judged next to this staggering low, President Barack Obama's administration gives us great reason for hope. But science remains marginalized in the political arena, and few elected officials really understand or appreciate its centrality to decision making and governance. Too many politicians, Democrats and Republicans alike, fail to see the underlying role of science in most of the issues they address, even though it is nearly always present. In fact, politicians tend to be leery of seeming too scientifically savvy: There's the danger of being seen as an Adlai Stevenson egghead.

We're still struggling with the problem that historian Richard Hofstadter outlined in his classic 1962 work, *Anti-Intellectualism in American Life*, which documented how the disdain of intellect became such a powerful fixture of American culture. The problem is particularly acute when it comes to scientists, and this has been the case to varying degrees since our nation's inception. We've even rewritten the biography of one of our most cherished founding fathers, Benjamin Franklin, recasting him as a tinkering everyman when in fact he was a deep-thinking scientist of the first rank. After visiting the country in the 1830s, Alexis de Tocqueville similarly remarked upon Americans' interest in the practical rather than the theoretical side of science, observing a people more intrigued with the goods delivered at the end of the process than the intellectual challenges and questioning encountered along the way. For a very long time, American scientists have found themselves pitted against both our businesslike, can-do attitudes and our piety. When John McCain and Sarah Palin ridiculed research on fruit flies and grizzly bears on the 2008 campaign trail, they were appealing to precisely this anti-intellectual strand in the American character. They thought they'd score points that way, and they probably did.

And if you think politicians are bad, let's turn to the traditional news media, where attention to science is in steep decline. A 2008 analysis by the Project for Excellence in Journalism found that if you tune in for five hours' worth of cable news, you will probably catch only one minute's coverage of science and technology—compared with ten minutes of "celebrity and entertainment," twelve minutes of "accidents and disasters," and "26 minutes or more of crime." As for newspapers, from 1989 to 2005 the number featuring weekly science or science-related sections shrank by nearly two-thirds, from ninety-five to thirty-four. These trends in both types of media have continued and perhaps even accelerated: In 2008, CNN shut down its science, space, technology, and environment unit, and in 2009, the *Boston Globe* killed its esteemed science section.

As a result of this upheaval, what we might broadly call science communication—the always problematic bridge between the experts and everybody else—is in a state of crisis. The business-driven cutbacks on science content by the "old" media are bad enough, but the "new" media are probably hurting science as much as helping it. The Internet has simultaneously become the best and the worst source of information on science. Yes, you can find great science content on the Web, but you can also find the most stunning misrepresentations and distortions. Without the Internet, the modern vaccine-skeptic movement probably wouldn't exist, at least not in its current form. Jenny McCarthy, celebrity vaccine critic extraordinaire, is proud of her degree from the "University of Google."

More generally, thanks to the Internet and ongoing changes in the traditional news industry, we increasingly live in an oversaturated media environment in which citizens happily try on information sources to see which fit them best. This means they can simply avoid science content altogether unless it seems a good personal match. And they can shop online for scientific "expertise" as easily as they can for Christmas gifts.

When we shift our attention to another extremely powerful source of information about science—the entertainment media—we find the situation more complex but still dismaying. From *Grey's Anatomy* to *CSI* to *The Day the Earth Stood Still* (the Keanu Reeves version), science and technology provide fodder for many popular television and film plotlines. In fact, there appears to be a growing trend of basing stories on scientific themes, especially in the case of prime-time medical dramas. But whether such entertainment depictions contribute to a science-friendly culture is less clear. Often we see little effort devoted to achieving basic scientific plausibility or getting the details right; and we simultaneously find Hollywood obsessed with paranormalist UFO and "fringe science" narratives and recurrent stories of "mad scientists" playing God. Scientists in film and television tend to be depicted as villains, geeks, or jerks. Rare indeed is the Hollywood film or scripted drama that tells a story about science that's both serious and entertaining. That strongly affects how we think.

And then there's religion, the source of the deepest fissure in the science-society relationship. Surveys overwhelmingly show that Americans care a great deal about faith; many scientists, by contrast, couldn't care less. There's nothing wrong with that, except that some scientists and science supporters have been driven to the point of outright combativeness by the so-called New Atheist movement, led by Sam Harris, Richard Dawkins, and others. Meanwhile, many U.S. religious believers are just as extreme: They reject bedrock scientific findings—including an entire field, evolutionary biology—because they wrongly consider such knowledge incompatible with faith. The zealots on both sides generate unending polarization, squeeze out the middle ground, and leave all too many Americans convinced that science poses a threat to their values and the upbringing of their children.

For all these reasons, the rift between science and mainstream American culture is growing ever wider. Nearly a decade into the twenty-first century, we have strong reason to worry that the serious appreciation of science could become confined to a small group of already dedicated elites, when it should be a value we all share.

Oddly, however, American scientists seem to be feeling pretty optimistic right now. They certainly feel much better than they did five years ago, when they began rallying in a fairly extraordinary fashion—especially for scientists, many of whom tend to view politics as something rather distasteful—to oppose the administration of George W. Bush.

The Bush administration featured unending scandals over political interference with science and scientists. The president himself misstated the facts about the number of embryonic stem cell lines that would be available for federally supported researchers, exaggerated scientific uncertainty about global warming, and kowtowed to anti-evolutionists. His political underlings, meanwhile, regularly gagged government scientists and rewrote their reports. In response to this incredible abuse, American scientists became strongly energized, denouncing the Bush "war on science" and eventually organizing into initiatives such as ScienceDebate2008, hoping to reform the way the political system treats scientific knowledge.

In this context, it's no wonder the Obama administration feels like salvation. Having a president who values science, who surrounds himself with experts and shows every indication of respecting what they tell him, who pledged in his inaugural address to restore science to its "rightful place" in our government—all this changes the cultural climate dramatically. It's reason to celebrate.

Yet we are deluding ourselves if we think all the problems surrounding science have suddenly been solved. If the Bush administration could become so outrageously anti-science, surely there must be something about our society that makes such behavior politically viable or advantageous—and easy to get away with. A change in administration doesn't automatically fix the underlying problems, which include the corporate media's marginalizing of science, ongoing divides over science and religion, and an American culture that all too often questions the value of intellect and even glorifies dumbness.

In fact, many observers of science policy fear that despite the best of intentions, the Obama administration could find its hands tied when it comes to advancing science in the long run. It will probably take most of the president's first term just to resolve some of the massive problems caused directly by our failure to take science seriously in recent years.

Consider the intertwined climate and energy issues. Scientific warnings about global warming go back decades, yet our political system has repeatedly failed to take action. We now find ourselves in a harsh predicament: Even if we move quickly to address the problem, some effects of global warming could still be devastating and irreversible. The only solution is to remake our energy economy, shifting fairly rapidly away from fossil fuels; but here again, our leaders have failed to adequately recognize the need for change, at least until relatively recently. U.S. research funding for energy innovation was in steep decline from 1980 to 2000 in both the corporate and government sectors, a staggering lack of foresight by both our representatives and the society that elects them.

It will require an unprecedented effort, but just maybe the Obama administration and the Democratic Congress can turn all of this around. In the process, we hope the president will continue to use the bully pulpit, as only he can, to explain to Americans the centrality of science to the solutions we must develop. But what of the next set of science-related issues, already visible ahead of us? They extend far beyond our admittedly massive climate and energy problem.

At a time of dramatic economic disruption, when scientific research has been a core driver of the nation's growth over the past century, U.S. government funding of research and development stunningly failed to keep pace with inflation for five years running between 2004 and 2008. Meanwhile, we watched other nations surge in scientific productivity and enthusiastically embrace science as the key to their futures. The American scientific community has been sounding the alarm about this competitiveness challenge, but the political sector has barely begun to respond. Thankfully, the economic stimulus package crafted by Congress and signed into law by President Obama in early 2009 was encouragingly generous to science and finally reversed the disturbing trend of funding declines. Yet we're not overly optimistic about longer-term funding prospects in a climate of trillion-dollar deficits. Fiscal priorities have a habit of shifting to immediate needs, and with our government trying to extinguish multiple fires at once, even the best-intentioned and most science-minded of administrations may have a hard time making truly visionary investments anytime soon.

Looking even further into the future, we can anticipate the coming controversies that new research, particularly in the brain sciences and genetics, could unleash. These days, science fiction is sounding a lot less fictive. Of course we can't fully predict the future, but it is already possible to anticipate some of what may be on the way: the creation of synthetic microbes in the laboratory; the artificial retardation of human aging; the birth of a generation of "designer babies"; the tailoring of medical treatments to our personal genotypes; the increasingly physical understanding of the workings of the brain and its role in individual actions, leading to all kinds of potentially troubling applications, such as the determination of guilt or innocence in the courtroom; and much more. We'll soon be discovering many new levers that could allow us to alter the nature of human identity and existence, and that is not the only kind of possible intervention the future may hold. We're also moving ever closer to the knowledge and techniques that will let us actively manipulate the planet's climate and weather—so-called geoengineering. Once we have this ability, and in truth we may already be there, won't we be sorely tempted to use it?

Having a scientifically attuned public and a scientifically infused culture will matter more than ever as divisive debates emerge about the propriety of such interferences with "nature." We ought already to be anticipating them and preparing for them as a society. But for the most part, we are not. Scientists know what advances are under way and debate them regularly at their conferences, but they're talking far too much among themselves and far too little to everybody else. This isn't a gap the president or his administration can bridge, and certainly not alone. We need the experts themselves to launch new initiatives to bring these topics into the spotlight, before it's too late to have a serious dialogue about them.

Let's not forget that even though the scientific community's old foes (anti-evolutionists, global warming deniers, and so on) may have fallen out of political power, they are no less determined. Moreover, they tend to be much more invested in cutting-edge communication and persuasion techniques than the defenders of science and reason. And they pull out all the stops when it comes to lobbying, argument framing, journalist arm-twisting, and just generally getting their views across, seizing upon a diverse array of media opportunities to do so.

If scientists don't find new ways of reaching out to the broader society in which they work, they should know all too well by now who will win the attention of the public, the media, and the politicians over the coming years.

It's not hard to understand why many scientists have been so reluctant to engage in such a battle. They still remember a time when keeping America focused on science seemed much easier. In the heady years following the Allied victory in World War II, American scientists enjoyed great cultural authority and access to the corridors of power, the invitation to rewrite the nation's educational curriculum, and much more. Many leaders of science still remember that era, yet often at their peril if they believe it reflects the natural or normal relationship between science and American society. Instead, this rapport requires tremendous effort to forge and maintain.

Our culture has changed vastly since the mid-twentieth century. Science has become much less cool, scientists have ceased to be role models, and kids aren't rushing home anymore to fire rockets from their backyards. It would be unproductive and also unfair to blame scientists alone for this sad state of affairs. For every scientist who shuns or misunderstands the broad public, there's another who deeply wants to find better ways to connect and who may exert considerable energy and ingenuity to that end. And we've already seen how other crucial sectors of society fail to give science its due.

Still, it is undeniable that the troubling disconnect between the scientific community and society stems partly from the nature of scientific training today, and from scientific culture generally. In some ways science has become self-isolating. The habits of specialization that have ensured so many research successes have also made it harder to connect outside the laboratory and the ivory tower. As a result, the scientific community simultaneously generates ever more valuable knowledge and yet also suffers declining influence and growing alienation. Too many smart, talented, influential people throughout our society don't see the centrality of science in their lives; and too many scientists don't know how to explain it to them.

We are not the first to diagnose the problem this way: Our argument has, as its patron saint, a scientifically trained British novelist named C. P. Snow. Fifty years ago, on May 7, 1959, Snow delivered a famous speech entitled "The Two Cultures and the Scientific Revolution." The scientists and humanists of his day, Snow lamented, not only failed to communicate; often they disdained one another. They stood separated by a "gulf of mutual incomprehension." And this wasn't a mere oddity of mid-century British intellectual life—it was a global phenomenon with grave consequences. "This polarization is sheer loss to us all," Snow stated. "To us as a people, and to our society."

Snow has often been accused of oversimplification with his "two cultures" thesis; as he himself admitted, "The number 2 is a very dangerous number." Yet Snow grasped one overarching truth: The rift between science and culture had to be mended. There were walls to knock down, gulfs to bridge, people to unite, and the future depended on it. Snow knew what really mattered, and you might say our book is merely here to provide half a century of transatlantic updating.

And to save Pluto, of course.

CHAPTER 2

Rethinking the Problem of Scientific Illiteracy

IF SCIENCE AND OUR CULTURE HAVE COME UNSTUCK, OR IF THEY NEVER properly adhered, we have a serious problem. But it's also one we need to think about in new ways. In this book we aim to show how science and American society have diverged sharply in the modern era, to describe the present state and consequences of this disconnect, and finally, to propose solutions. First, however, we must dispel some prevalent misconceptions about the real nature of the problem and who is responsible for its existence.

Among many scientists, there have long been groans about the public's "scientific illiteracy." The evidence usually consists of various embarrassing survey findings, revealing disastrously poor citizen responses to questions about scientific topics they presumably studied in elementary or high school. (For instance: "Electrons are smaller than atoms, true or false" or "The universe began with a huge explosion, true or false.") One prominent researcher on the public understanding of science has even found that due to their failure to understand basic scientific terms or the nature of the scientific process, 80 percent of Americans can't read the *New York Times* science section. Perhaps the most shocking and oft-cited scientific illiteracy result: Only half of the adult populace knows the earth orbits the sun once per year.

Such dismal findings have given rise to a standard complaint about where the problem lies whenever scientists and our society, or our political system, come into conflict. The blame is said to lie with "the public," which needs to be more educated, more knowledgeable, better informed. Yet even a cursory examination reveals serious problems with this line of thinking.

To begin with, citizens of other nations don't fare much better on scientific literacy surveys, and in many cases fare worse. Residents of the European Union, for instance, are less scientifically literate overall than Americans, at least according to one metric for measuring "civic science literacy" across countries. And yet they also appear much more convinced of the reality of global warming and human evolution.

Such complexities call into question whether quizzes about a few canonical "facts" or the nature of the scientific process really tell us much about a society's outlook on the science issues that matter most. Indeed, it's doubtful that a baseline level of scientific literacy is remotely adequate for engaging with the science-centered debates that play out regularly in the news media and the political arena. Is the goal to have a public that can dig into complicated scientific disputes and determine who is right or wrong? If so, then let's remember that many anti-evolutionists and global warming deniers are scientists themselves, couching their claims in sophisticated scientific language and regularly citing published articles in the peer-reviewed literature. To refute their arguments, one often needs Ph.D.-level knowledge. And even then, the task requires considerable research and intellectual labor well beyond the resources or interest of most people.

And the problem grows even more complicated, because sometimes those citizens who put in the most work to understand scientific topics come out the very worst in the end—more severely *mis*informed than if they were merely ignorant. As Mark Twain put it, "The trouble with the world is not that people know too little, it's that they know so many things that just aren't so." Take the army of aggrieved parents nationwide who swear vaccines are the reason their children developed autism and who seem impossible to convince otherwise. Scientific research has soundly refuted this contention, but every time a new study comes out on the subject, the parents and their supporters have a "scientific" answer that allows them to retain their beliefs. Where do they get their "science" from? From the Internet, celebrities, other parents, and a few non-mainstream researchers and doctors who continue to challenge the scientific consensus, all of which forms a self-reinforcing echo chamber of misinformation.

The vacine-autism advocates are scientifically incorrect; there's little doubt of that at this point. But whether they could be called "ignorant" or "scientifically illiterate" is less clear. After all, they've probably done far more independent research about a scientific topic that interests and affects them than most Americans have.

The same goes for other highly informed, and deeply wrong, groups—the global warming deniers, anti-evolutionists, UFO obsessives, and so on. Ignorance isn't their problem, and neither is a lack of intellectual engagement or motivation. Anyone who has ever discussed global warming on national radio—as Chris has done countless times—can expect to be besieged by callers who don't accept the prevailing scientific consensus and have obviously done a great deal of research to back up their prejudices. If anything, such individuals want to make a show of their erudition and proceed to rattle off a mind-boggling string of scientific-sounding claims: Global warming isn't happening on other planets; urban heat islands (cities) thwart global thermometer readings; the atmosphere's lowest layer, the troposphere, isn't warming at the rate predicted by climate models; and the like.

Or consider the late Michael Crichton. He was a brilliant science-fiction novelist, screenwriter, and movie producer who backed up his best-selling narratives with considerable scientific research. Yet in his late-life novel *State of Fear*, he penned a wholly misleading and revisionist attack on the science of global warming. Faced with such people, intellectually driven and empowered as never before by the profusion of "science"—good, bad, and awful—on the Internet, one soon recognizes that the lack of scientific knowledge probably isn't our main problem.

Almost inevitably, improvements to our educational system are put forward as the primary solution to the problem of scientific illiteracy. It is a lofty goal, of course, and nobody is *against* improving K–12 science education. But to look to education alone as the silver bullet is to write off as unreachable anyone who has already graduated from the formal educational system. That includes vast stretches of the population, including most voters, our political and cultural leaders, and the gatekeepers of the media.

The most troubling problem with the standard "scientific illiteracy" argument, however, is this: It has the effect, intended or otherwise, of exempting the smart people—the scientists—from any responsibility for ensuring that our society really does take their knowledge seriously and uses it wisely. It's an educational problem, they can say, or a problem with the media (which doesn't cover science accurately or pay it enough attention), and then go back to their labs.

The Pluto saga, which captured vastly more attention than most science news stories ever do and deeply engaged many members of the public, utterly explodes this conceit. There isn't any obvious "true" or "false" answer to the question of whether Pluto is a planet, and people certainly weren't ignorant about it. Rather, they were outraged by the sudden, top-down, seemingly arbitrary change by the science world, and they weren't necessarily wrong to have that reaction.

For all these reasons, scholars working in the field of science and technology studies (STS) have largely discarded the idea that our problems at the science-society interface reduce to a simple matter of scientific illiteracy, traditionally defined. Instead, these thinkers have grown skeptical of what they sometimes call the "deficit model" that has come to dominate many scientists' and intellectuals' views of the public—the idea that there's something lacking in people's understanding or appreciation of science, and that this in turn explains our predicament.

The "deficit" outlook usually takes a benign form, casting scientists in the role of benevolent tutors to a public starved for knowledge. But it can also turn nastier, morphing into what we might call the "you're an idiot" model. All too often we find scientists saying things to their peers and colleagues, or even to the press, that sound something like this: "I can't believe the public is so stupid that it believes X" or "I can't believe people are so ignorant that they'll accept Y." At this point the scientist ceases to be a friendly instructor and becomes a condescending detractor and belittler.

Either way, the "deficit" approach fails to offer effective ways of reaching people with accurate scientific information and making it stick. Members of the public aren't empty vessels waiting to be filled with science; the refusal to tailor such information to their needs virtually ensures it won't be received or accepted. And pointing fingers at the public or its surrogates—politicians, journalists, celebrities, and so on—is not only insulting and alienating but discourages reflection about the role scientists might be playing in the equation. Perhaps most troubling, as science-communication scholars have noted, the finger-pointing approach can trigger a vicious circle:

> A deficient public cannot be trusted. Mistrust on the part of scientific actors is returned in kind by the public. Negative public attitudes, revealed in large-scale surveys, confirm the assumptions of scientists: a deficient public is not to be trusted.

So although we share with scientists the concern that their work isn't adequately appreciated or heeded in our culture, this book will not unfold as a litany of all the ways in which the public falls short in its scientific knowledge. Neither will we proceed by exposing all the nonsense that people are regularly fed in place of good science: quack alternative medicine claims, fringe attacks on mainstream environmental research, paranormal obsessions, and the like. We're more interested in divides and how to bridge them.

That's not to say, however, that we wish to entirely discard the concept of "scientific illiteracy." We'd prefer to redefine it, getting past issues of finger-pointing and buck-passing and the misconception that our problems can be reduced to what non-scientists say in response to survey questions.

Luckily, there's another side to the scientific literacy tradition, one that goes beyond the standard emphasis on factual or theoretical scientific knowledge to stress a third aspect: citizens' awareness of the importance of science to politics, policy, and our collective future. This dimension has often fallen by the wayside in debates about scientific illiteracy, and yet we believe it is easily the most important.

In this sense, there's scarcely any doubt that we are scientifically illiterate, and dangerously so. But the problem is with our society as a whole, and we all sink or swim together. It's not just the fault of the non-scientist public or the educational system: Scientists, who as a group have come to share assumptions, practices, and behaviors that place them at a far remove from their fellow citizens, play a crucial role in this dynamic. To use a classroom analogy, if the students are throwing spitballs and paper airplanes, the teacher is also droning on interminably and barely seems to notice. For this unproductive chaos we all share responsibility, scientists and non-scientists alike.

And anyway, we don't need average citizens to become robotic memorizers of scientific facts or regular readers of the technical scientific literature. Rather, we need a nation in which science has far more *prominence* in politics and the media, far more *relevance* to the life of every American, far more *intersections* with other walks of life, and ultimately, far more *influence* where it truly matters—namely, in setting the agenda for the future as far out as we can possibly glimpse it. That would be a *scientific America*, and its citizens would be as scientifically literate as anyone could reasonably hope for. We will never have a nation that is fully composed of Ph.D.s.

This future-oriented perspective also helps us see why having a society shot through with scientific illiteracy poses such a threat. It leaves us too little attuned to the fundamental advances and dynamics that will inevitably shape the coming decades. The result is our repeated failure as a nation to take forward-looking actions before it's too late.

Having untangled the concept of "scientific illiteracy," we can move on to dispel another misconception: the idea that the American public is "anti-science" in any meaningful sense of the term.

Polling data refute this notion outright. Rather than actively disliking science, Americans have at least some positive attitudes toward it. For instance, America's scientific leaders still enjoy more public confidence than the leaders of any prominent institution other than the military. Still, the vast bulk of Americans cannot even name those scientific leaders they so trust, and that points to the real nature of the difficulty we face.

It's not that most Americans despise science. Rather, they're too uninvolved; they don't have science on the radar most of the time. When directly asked in surveys, members of the public express considerable interest in learning about new scientific discoveries. Yet clever pollsters have been able to tease out that these respondents are far *more* interested in other things. According to the National Science Foundation, only 15 percent of the public follows science news "very closely," meaning that science ranks behind ten other news subjects, including crime, sports, and religion, in its ability to hold people's interest. (Science's ranking vis-à-vis other news topics has been slipping of late, and declining treatment of science in the news media reflects as much.)

We fully concede that it could be worse: Americans could actively hate science. But the public's highly superficial degree of appreciation, forgotten at a moment's notice, won't suffice for what we face as a nation. The failure to recognize the importance of science now will hurt us in the decades ahead, especially economically, and will leave us unprepared for the controversies and challenges already on our doorstep. Meanwhile, disengagement from the vibrant world of science leaves our citizens all too susceptible to rampant misinformation, inaccurate anti-scientist stereotypes (the socially challenged geek, the arrogant madman), and the anti-intellectual tendencies that have plagued our national character for too long.

And yet recognizing that matters are far from hopeless should be strongly empowering: We can do better. As the great historian Richard Hofstadter explained, during different periods Americans have lurched closer toward, and further away from, strong anti-intellectualism. The transition from the Bush to the Obama administration represents a perfect example. Similarly, the public seems capable of going either way on matters of science, and we can certainly move people toward a broader acceptance and appreciation of its centrality to the future. Yet we must recognize that many forces stand arrayed against this necessary project, perhaps most centrally, the ongoing convulsions in the modern media. And thus far the science world and its allies have not taken adequate steps to counter them.

So what can be done? The first step is to understand our history and unravel how we reached a point where America's scientists dazzle the world, yet at home encounter a public that too often shrugs its shoulders. To that end, the first section of this book traces the rise and relative decline of science's political and cultural standing in the United States since World War II, with a central focus on the troubled and often halting attempts by scientists to reach out to the broader public and the changing societal factors (such as the transformation of politics and the media) that have made this prospect increasingly difficult. Going over this ground will help us shift away from a blame-oriented analysis of the science-society gap—which targets the public, the media, the politicians, or the educational system—toward something more fruitful.

But if we're discarding the "deficit model," what can take its place? Historical awareness can help us here as well. C. P. Snow's "two cultures" argument has great potential to reinvigorate our discussions of the science-society disconnect in America and shift us away from the problems inherent in deficit thinking. Approaching our science-society problems through Snow's "mutual incomprehension" framework does not require casting anyone in the role of superior or inferior, smarter or dumber.

If it is to serve this purpose, however, Snow's argument will first require a few renovations. In his era in Britain, the culture of arts and letters enjoyed preeminence; science, by contrast, was the underdog. Now things have flipped, at least within the academic arena, even as that arena has become conspicuously less influential overall. So when we discuss Snow's ideas today, we must go beyond simply lamenting a divide between scientists and humanists. Yet the core of Snow's concern—that science isn't being translated broadly enough into societal and political relevance, and that this has something to do with too much specialization and compartmentalization of knowledge—remains as valid as it ever was. The same goes for the "two cultures" concept, so long as we're willing to perform some simple multiplication.

Today in the United States we find science walled off, in a classic Snowean sense, from not one but many "cultures" that, together, powerfully shape the course of our thinking—most notably political culture, media culture, entertainment culture, and religious culture. The second section of this book will therefore discuss these four major rifts in sequence and begin to propose ways of bridging them. That's not to say the science-humanities divide that so worried Snow has gone away; it hasn't. However, its importance has been dwarfed by divides between science and sectors that aren't really even part of the academy at all. This development poses new and very different challenges to scientists, who tend to be most at home in a university setting (even if they don't always see eye to eye with their fellow scholars).

If our analysis is correct, it follows that the problem really isn't that Americans cannot recite enough scientific facts or that the smart people are mobbed by idiots, but rather that we have far too many unhealthy disconnects between different types of talented, intellectually motivated leaders and thinkers. There are too few collaborations between scientists and journalists, screenwriters, politicians, and religious leaders. The goal must be to remake our educational system and our cultures, both scientific and popular, to generate much more interplay among different kinds of talent and expertise. At the same time, we must rouse the people who care about science and inspire them to reach out to other parts of society, and to the public as a whole, in order to openly engage rather than criticize or blame.

To that end, we must stoke an ongoing cultural change at the nation's leading scientific institutions and universities. Even as they train the next generation of scientists to help keep us competitive in the global economy, these centers of science must reward endeavors that they have undervalued for far too long: public outreach, communication, and reuniting the "two cultures" through true interdisciplinary education (which must go far beyond exploring, say, the intersections between biology and chemistry). The final section of this book therefore looks to the changing media environment and changing university culture to explore how we can ring in the reforms, as well as the new movements and attitudes, that we're going to need. The ultimate solution may require nothing short of redefining the role of the scientist in today's society. But if that will make the society itself more scientifically engaged, surely it would be worth it.

PART I

THE RISE AND CULTURAL DECLINE OF AMERICAN SCIENCE

Dr. Hathaway: Mitch, there's something you're going to have to understand. Compared to you, most people have the IQ of a carrot.

—*REAL GENIUS*, 1985

CHAPTER 3

From Sputnik to Sagan

THE MODERN HISTORY OF AMERICAN SCIENCE BEGINS WITH WORLD WAR II. The conflict created plenty of heroes, and among them stood scientists themselves, who could claim credit for developing crucial wartime technologies such as radar and the atomic bomb. As one contemporary scientist wrote, after the war "suddenly physicists were exhibited as lions at Washington tea parties." They had saved the free world. They were superstars.

Postwar, the scientific community found itself swept up in the national mission, and scientists came to enjoy considerable cultural and political authority. In 1944, President Franklin D. Roosevelt asked Vannevar Bush, the president of the Carnegie Institution of Washington and director of the wartime Office of Research and Development, to undertake a study of how the institutions of science, having so impressively proved themselves, could continue to serve the nation in the "days of peace ahead." Bush and Roosevelt shared a mission and also a powerful metaphor, one that closely aligned science with the national identity. "The pioneer spirit is still vigorous within this nation," wrote Bush in a cover letter addressing the president's request. "Science offers a largely unexplored hinterland for the pioneer who has the tools for his task."

Bush's report, entitled *Science: The Endless Frontier*, extended the metaphor and laid out what would become the postwar consensus on the centrality of science to America's mission. The government, the report argued, should invest heavily in basic scientific research conducted at U.S. universities, and in turn, the knowledge produced would lead to technological advances that would enrich our lives—improving health and medicine, spurring economic growth and the creation of jobs, and strengthening the national defense. A generation of "scientific talent" would be trained in the process.

Some details aside, the vision Bush outlined in his report largely came to pass. Prior to World War II, the U.S. government had invested in some discrete areas of science, such as agriculture, but had no broad or systematic approach to funding research. Independent entrepreneurs—such as Thomas Edison and the Wright brothers—were by and large driving innovation in the early part of the twentieth century. Government parsimony wasn't the only problem: Before World War II, many institutions of American science remained leery about accepting government funds even if they were available, fearing the integrity of their work would be compromised. (European scientists throughout this period not only enjoyed greater fame and productivity but showed far greater willingness to accept the patronage of government.)

Thanks to the war, all that changed. Many top European researchers, fleeing Hitler's and Stalin's armies, had now emigrated to the United States. And the institutions of American science began to receive considerable amounts of government money. From 1953 to 1961, federal funding for research and development grew by 14 percent per year.

In particular, the Soviet launch of Sputnik in 1957 triggered a huge spike in government funding for the sciences. The first Earth-orbiting satellite, beeping at us from above, inspired stark fears about our national security and competitiveness: Were we falling behind in technology? Would the Soviets fire on us from the skies, and if they tried, could we stop them? As Senator Lister Hill, an Alabama Democrat, put it, the nation had experienced "a severe blow, some would say a disastrous blow, at America's self-confidence and at inner prestige in the world." If the Soviets beat us to the moon, added sci-fi visionary Arthur C. Clarke, "they will have won the solar system, and theirs will be the voice of the future . . . As it will deserve to be."

In response, the U.S. Congress jacked up the budget of the recently formed National Science Foundation to $134 million, an increase of nearly $100 million in just one year. Over the next decade the explosive growth continued, so that by 1968 the NSF budget approached $500 million. And the NSF represented just one slice of the science pie: Many other agencies, especially military ones, were also funding research. In 1940, the total federal R&D expenditure stood at around $74 million; by 1962–1963, it had grown to $12.2 *billion.*

The political and cultural dedication to science didn't limit itself to research dollars. In 1958, Congress also created NASA, home to America's moon-bound space program, and passed the National Defense Education Act, providing generous funding to encourage American students to pursue careers in science and engineering. Graduate fellowships, low-interest college loans, and new research equipment all flowed from the government, to the tune of $1 billion over the course of four years. Elementary and high school science education also received dramatic new attention, as science-intensive curricula, funded and championed by the National Science Foundation and designed by the nation's scientific elite, swept into the public schools.

The launch of Sputnik also led to a much closer integration of scientific expertise and political decision making. President Dwight D. Eisenhower answered this national science challenge by pulling the country's top researchers into his inner circle and seeking their unfiltered advice. In 1957, he appointed the first official presidential science adviser, MIT president James Killian, and launched the distinguished President's Science Advisory Committee. As Eisenhower would later put it, "This bunch of scientists was one of the few groups that I encountered in Washington who seemed to be there to help the country and not to help themselves."

The wedding of science and policy continued under President John F. Kennedy. He too adopted the "frontier" metaphor to talk about scientific discovery, and through such inspirational language stoked the conviction that Americans could put a man on the moon. The Apollo program, begun on his watch, ultimately fulfilled that vision. Kennedy also bulked up the science advisory apparatus further by giving his own adviser, MIT electrical engineer Jerome Wiesner, a new White House office to run, the Office of Science and Technology.

Perhaps C. P. Snow himself best captured the science-euphoric mood of the time. In a 1960 lecture before the American Association for the Advancement of Science (AAAS), the leading general membership organization for the science community in this country, Snow remarked that "scientists are the most important occupational group in the world today. At this moment, what they do is of passionate concern to the whole of human society." During the 1950s and early 1960s, much of America seemed to agree.

Yet the heyday of science wouldn't last long. The trend of ever-rising federal investment in research reversed itself in the late 1960s; non-military science funding fell through much of the 1970s. The central involvement of the National Science Foundation in shaping high school educational curricula also gradually fell away—as did the prominence of the scientific elite in advising our leaders.

All this occurred for a snarl of reasons that center on the collapse of political consensus in America over the same time period. The scientific community couldn't escape the conflicts of the day any more than any other major social group. The creation of new regulatory agencies, like the Environmental Protection Agency in 1970, repeatedly dragged scientific information into a contested decision-making process. And the growing antiwar and antinuclear movements flailed not only against the "military-industrial complex" but against those parts of the scientific establishment that worked with it or for it, especially on university campuses.

Not only did the new mood of "questioning authority" include the questioning of science, but there was often good reason for skepticism. The environmental and consumer movements, spearheaded by the likes of Rachel Carson and Ralph Nader, brought home the realization that science wasn't always beneficial. Seemingly wonderful technologies— DDT, chlorofluorocarbons—could have nasty, unforeseen consequences. A narrative began to emerge about "corporate science": driven by greed, conducted without adequate safeguards, placing profits over people.

Amid the backlash, the role of scientists in policy making also ebbed, as did the status of the presidential science adviser. Whereas the post had enjoyed a high level of recognition and influence under Eisenhower and Kennedy, the Vietnam War brought that cozy relationship to an end. In 1973, President Richard Nixon fired his science advisers outright over disagreements about the viability of the Supersonic Transport program and other matters.

The emergence of the Religious Right onto the political stage in the 1970s—motivated in part by its adherents' resentment of the nation's intellectual and scientific elites—was also a major factor in curtailing the role of science in public policy. Soon battles between Christian conservatives and the science community over matters like the teaching of evolution and embryo-related research became an inescapable political reality. And so we entered the "culture wars": Secular, scientific, and pro-choice America clashed regularly with a "faith-based" (and very Republican) side of the country. A vast array of political issues could be cast in a pugilistic context that pitted "religion" against "reason." Science ceased to serve as a bulwark for common goals and purposes; instead, its findings came to divide us.

These were big, momentous changes, and it's not as if scientists didn't notice or respond to them. But there was always a tension: Their wartime role and post-Sputnik glory notwithstanding, many scientists simply wanted to stay out of politics and get on with fundamental research. They preferred being left alone to pursue the expansion of knowledge, with new infusions of financial support but few other commitments or requirements. As British government science adviser Sir Solly Zuckerman described the mentality in 1970:

> What the pure scientist basically wants and needs is the assurance that he will be allowed to give full rein to his curiosity without being harried, until the moment comes when he himself thinks his ideas have either flowered or run into the sands, and it is time to change direction, or give up research.

In his 1967 book *The Politics of Pure Science*, the journalist Daniel Greenberg launched a grenade into the scientific academy by targeting precisely this mindset. Greenberg criticized scientists' devotion to the value of pure science or so-called basic research—science generally cultivated within university settings and directed at extending knowledge rather than in service of any immediate technological application—above all else. The problem lay in what Vannevar Bush had said about basic research: It was performed "without thought of practical ends." New technologies might ultimately grow out of it, but not in an immediately foreseeable way.

So those scientists considered to be doing the most important work were very often doing research that non-scientists and politicians—who do not even grasp the basic distinction between pure research and the pursuit of new technologies in many cases—would find the least accessible or even esoteric. As Greenberg put it, "Modern science not only is inscrutable to the masses, but, in many of its manifestations, has become progressively dissociated from humankind." Another closely related phenomenon worsened matters: academic specialization. Whereas in 1945, scientific journals numbered somewhere between 5,000 and 10,000, by the 1990s that number had risen to close to 100,000. And within their respective fields, scientists developed specialized jargons that impaired communication not only with the public but even with other scientists. By 1999, the late science popularizer and evolutionary thinker Stephen Jay Gould could lament:

We have now reached the point where most technical literature not only falls outside the possibility of public comprehension but also (as we would all admit in honest moments) outside of our own competence in scientific disciplines far removed from our personal expertise.

As the nation's consensus about the role of science in creating a better life for all Americans came into question, then, scientists did relatively little to counter the trend. Although the scientific community made some attempts to reach out to a broader public in the second half of the twentieth century, it's fair to say these efforts were ambivalent and often tentative. Public engagement would always take a backseat to research, as leading statesmen in science fully acknowledge: "For 45 years or so, we didn't suggest that it was very important for all these scientists we're talking about to invest much of their time [explaining their work to the public]," National Science Foundation director (and later, President Bill Clinton's science adviser) Neal Lane observed in 1997.

> In fact, we said quite the other thing. We said: what's critically important, since we are investing taxpayer money for discovery of new ideas about nature and new ways of doing business and new kinds of devices to help humankind, we want you to be in the laboratory, in the classroom— that's what you're capable of doing and that's really where you ought to spend your time.

Spend it they did. Science's achievements over these decades are stunning, from landing a spacecraft on our own moon to landing a robotic probe on Saturn's; from discovering the double-helical structure of DNA to sequencing the human genome. But as they homed in on research, American scientists largely relied on two key surrogates— educators and journalists—to do their public relations for them.

Scientists have long championed public science education; in the late 1950s and early 1960s, they even took a central role in trying to shape it. But the curriculum reforms pushed so strongly by the nation's top scientists, with National Science Foundation support and funding, suffered from what we would now recognize as fairly flagrant "deficit" assumptions, and in any case, they ultimately achieved inadequate penetration into schools. In light of the huge ambition of these reforms, which sought nothing less than to create a science-centered America, it is depressing to note that over the thirty-year period from 1969 to 1999, U.S. science education faltered and students' science test scores remained "stagnant," to quote the National Academy of Sciences (NAS).

Beyond education, scientists could also look to a group of delegates to evangelize publicly on their behalf. These were the science journalists, who emerged in the first half of the twentieth century to communicate the excitement of discovery to a broader audience. The National Association of Science Writers, originally formed in 1934, was a small group at first but burgeoned during and after World War II. In many ways this new breed of writer served in part as a science publicist. Adoring science reporters worked closely with scientists to communicate the import of their research to the general public. Unlike today, it was a time when newspapers were increasing their quota of science journalism, and when scientists professed themselves happy with the general tenor of science coverage in the media. In one contemporary survey, 86 percent of them called newspaper science stories "reasonably accurate."

Yet even in those days, it was never clear how popular "popular science" could actually become. One striking indicator of the problem came in 1949, when the popular-science magazine *Science Illustrated* fell into disrepute after it started putting scantily clad women on its cover. In contrast, *Scientific American*, which catered more closely to scientists themselves, became profitable within three years of its relaunch in 1948. Science journalism was finding an audience, but it seemed a limited and circumscribed one—"preaching to the converted," as one science-communication scholar put it.

In sum, then, scientists were focused on basic research and growing increasingly specialized; high school science students weren't progressing; and science media attempts were revealing their limitations. Meanwhile, the cultural heyday of science was waning, and the political climate was growing vastly more challenging.

Such was the situation when a true science popularizer emerged and began to reach the broadest public conceivable, shattering boundaries and demonstrating what was really possible. But while America loved him, many other scientists responded with scorn or envy; and his popularity surge largely failed to lift other science-communication efforts for the long term. He represents, essentially, an exception, a missed opportunity, and an innovation the science world wasn't quite ready for, and didn't fully appreciate or welcome.

Carl Sagan entered the field of astronomy at the best of possible times for doing so: the 1950s and early 1960s. Government money was pouring into the space sciences. Even before he turned thirty, the charismatic and deeply self-assured young researcher found himself funded by the military, consorting with Nobel laureates, and publishing in the esteemed journal *Science*. But Sagan was a great deal more than a researcher. He was a skilled communicator, a master at connecting with ordinary people and explaining complicated science in terms they could understand.

From the start, Sagan was mediagenic and larger than life. But the true goad toward celebrity came in the mid-1970s, when he found himself frustrated by lackluster media coverage of NASA's *Viking* missions to Mars. Sagan decided to do journalists' work for them, and soon he and a colleague came up with the idea for what would become *Cosmos*, "the greatest media work in popular science of all time," as Sagan's friend Stephen Jay Gould later called it.

First aired on PBS in 1980, the series won an Emmy and a Peabody award, reached an estimated 500 million viewers around the globe, and galvanized untold numbers of students into scientific careers. The book version of *Cosmos*, also published in 1980, sold more than a million copies and further established Sagan's status as the greatest science popularizer in a generation. Unafraid of the mainstream media, Sagan kept science before the public eye with regular appearances on *The Tonight Show* with Johnny Carson, which featured him as a guest roughly twice per year over the span of more than a decade. He also managed to write or coauthor nearly twenty other books, one of which, his 1985 novel *Contact*, became a Hollywood blockbuster that grossed $170 million at the box office, and one of which, 1977's *The Dragons of Eden*, won the 1978 Pulitzer Prize for general nonfiction.

Sagan owed his incredible popularity to a unique set of talents. He was smart, attractive, confident, well-spoken, and an artful writer. But he was also a bold generalist, a multidisciplinary scientist who ranged across a wide variety of subjects. Throughout his career, he resisted academic specialization. The topics that fired him up weren't narrow or technical, but grand and imaginative: Did extraterrestrial life exist? What would it look like? Could we contact it, or it us? And what did the knowledge gained from science say about our place in the universe and imply for our religious faith?

The 1980s were Sagan's decade. In their first October, at the height of his *Cosmos*-inspired fame, he appeared on the cover of *Time* magazine, dressed in his characteristic turtleneck and standing on a shoreline with waves lapping at his legs. "Showman of Science," the cover said. Inside, the lengthy cover story, by science writer Frederic Golden, positively gushed. With Sagan as the "pathsetter"—the "prince of popularizers, the nation's scientific mentor to the masses"—science was becoming cool again. The disillusionment of the late 1960s and 1970s was wearing off; "ennui" had "turned into enthusiasm"; "eventually, the awe of science overcame the indifference to it."

The market appeared to agree. Between 1977 and 1984, a rejuvenated popular-science movement generated fifteen new magazines, eighteen newspaper sections, and seventeen television shows. The new ventures included CBS's *Walter Cronkite's Universe* and two major magazines, *Discover* and *Science 80*, which by 1983 had joined *Science Digest* in a triumvirate of mass market publications whose circulations would reach between 500,000 and 1 million each—at least temporarily. Science lovers got so giddy that one termed such publications "the general interest magazines of a new age."

Yet within just a few years, the sense of momentum had vanished. CBS canceled Cronkite's *Universe* in 1982—the same year ABC shut down *Omni*—and soon the bloodletting spread to the magazines. In 1986, *Science Digest* and *Science 86* both folded in the face of declining advertising revenue, and previous claims of a popular-science "boom" shifted to cries of a bust. Not everyone was a Carl Sagan, it appeared; the publishing and media market could be a harsh judge of otherwise well-meaning efforts.

Although it seemed that economic forces were chiefly responsible for this collapse, some charged that the leaders of the scientific community also deserved a share of the blame. Consider the case of *Science 80*—or *81*, or *82*, or *86*, depending on the year—published by the American Association for the Advancement of Science. Rather than helping prop up the struggling magazine as its advertisers pulled back, the AAAS sold out to Time, Inc., which killed it. Scientists have "declared that they don't want to be a part of" communicating with the public, one *Science 86* staffer angrily charged. "Presumably, they feel they have more important things to do. They don't." Indeed, even as the science media bubble was bursting, a nationwide survey of researchers found considerable institutional barriers to the popularization of science. The scientists surveyed agreed that "there is little to be gained within science by engaging in the public dissemination of information."

All of which is rather odd, because reconnecting with the American public was getting to be a pretty urgent matter for the scientific community. As early as 1980, the noted social scientist and public opinion expert Daniel Yankelovich could be found ominously predicting that the coming decade would be a "societal cockfight, with science and technology in the middle," and worrying loudly about the gap between science and what he termed the "public process."

As they watched the popular-science "bust," however, some scientists and science supporters seemed to feel less like redoubling their efforts, and more like giving up. Gerald Piel, the publisher of *Scientific*

American (which was also struggling), told science-communication scholar Bruce Lewenstein at the time that "the task of crusading for the popularizing of science to large audiences has to await better education in our schools," essentially arguing that popularizers should wait (for a decade or more) for a newer, smarter generation of science lovers. Piel wasn't the only one preaching defeatism. In 1981, the former science writer and Purdue University communication professor Leon Tracht-man put the case for retreat far more persuasively in a much-cited essay in the journal *Science, Technology, and Human Values.* Question-ing all the "missionary activity" on behalf of science, Trachtman asked whether anyone had considered the risks. More media coverage of sci-ence, he suggested, could actually lead to declining scientific credibil-ity with the public as reported findings were overturned or reversed by subsequent studies:

> It is difficult to communicate to the public the actual tentative, prob-ing, frequently intuitive nature of much of science. Instead, the public image of science tends to be one of a methodological force, ruthless and unstoppable in its logical and rational assault on the problems that face mankind . . . Holding this popular misconception, however, the public may develop expectations of science which science, by its very nature, cannot satisfy and which may, therefore, result in great public disillusionment.

Trachtman raised important concerns, but he probably took the pes-simism to an extreme; he even confessed himself to be acting in part as an *"advocatus diaboli."* And it wasn't exactly the best time to be flagging the *dangers* of popularization. Politically speaking, America was enter-ing the worst decade for science in the postwar era, a period that would set alarming precedents for the coming decades. The cultural rivals of America's scientific community—religious conservatives—were soon to get their first real taste of power. And the president of the United States, who owed his election in significant part to their mobilization, would undermine scientific knowledge and expertise repeatedly, including on matters at the center of his policy agenda, such as national defense.

As perhaps the chief public face of American science during this period, Carl Sagan wasn't merely a popularizer but a fierce advocate for the proper use of science in the real world. During the 1980s, President Ronald Reagan necessarily became his chief foe, for Reagan brought anti-science into the American political mainstream as never before. His worst abuses: smiling on creationism (just as he had as the governor of California) and making Star Wars—a sci-fi fantasy—the center of his foreign policy. Top physicists were baffled at the notion that so-called bomb-pumped X-ray lasers, suspended in space, could shoot down enemy ballistic missiles so flawlessly that America would be safe in the event of nuclear war. Each time a group of outside experts took a look at Star Wars (technically, the Strategic Defense Initiative, or SDI), they voiced grave doubts as to its technological feasibility as well as its strategic wisdom. Yet Reagan's policy, announced in 1983, went forward anyway, without anything resembling serious scientific vetting.

When Sagan first heard that Reagan had given a surprise speech announcing a new way to make nuclear weapons "impotent and obsolete," he was in the hospital, dealing with the increasingly difficult health conditions that would eventually end his life in 1996. But even though he was still on life support in the wake of surgery, Sagan called over his wife, Ann Druyan, to help him rally top scientists behind an anti-SDI petition.

Star Wars was just one of Sagan's targets. He also challenged the Reagan administration by publicizing his "nuclear winter" hypothesis, which countered the hawks' assumption that a nuclear war could, in any sense, be "winnable." Using computer simulations, Sagan and a group of other scientists found that in a situation of nuclear fallout, the fires produced could create so much smoke and dust that they might block out the sun's rays, creating a sustained planetary cooling that could threaten agriculture and possibly trigger global famine.

With all of this, Sagan greatly enraged the right wing. He became the target of attacks from William F. Buckley Jr.'s *National Review*, and from Buckley himself. He stirred up a hornet's nest of Reaganite, cold-warrior scientists, many of them centered on a newly formed conservative think tank called the George C. Marshall Institute. These scientists clashed repeatedly with Sagan in public debate as they defended SDI and lampooned "nuclear winter." But Sagan remained resolute. Invited to the White House three times by the president, who no doubt hoped to temper the attacks with hospitality, Sagan refused each time—and instead got himself arrested protesting at a Nevada nuclear test site. He was carrying the mantle of Albert Einstein, another scientist who, having achieved an extremely high level of public visibility, then reinvested that political capital into a core set of causes.

And like Einstein, Sagan was incredibly effective. The Reagan administration couldn't brush him aside the way the George W. Bush White House later did with the many distinguished scientists who organized to denounce his policies: Sagan was too famous and too well connected. He had regular access to the mass media as well as many world leaders, including the pope and Soviet leader Mikhail Gorbachev. In the end, Sagan played a significant part—hardly the only part, but a significant one—in helping to usher in a thawing of relations with the Soviet Union and the arms reductions that followed.

Sagan's battles during the 1980s exposed the growing anti-scientific bent of the Republican Party. The Star Wars and nuclear winter fights served as a kind of initiation rite for many conservatives who would develop ever more sophisticated strategies for attacking and countering mainstream scientific experts, like Sagan, in the future. The George C. Marshall Institute, originally launched by Sagan foe Robert Jastrow, would morph in the 1990s into one of the top outlets for sowing doubts about global warming. Scores of other think tanks with similar missions and methodologies would join it.

And Reagan's administration didn't just harm science on a substantive level. It also launched the trend of media deregulation that would, over time, help create an industry in which serious science coverage—including programs like *Cosmos*—had an increasingly difficult time surviving. Under Reagan, the Federal Communications Commission relaxed media ownership rules, opening the door to greater concentration of media outlets in fewer hands, and decreased public-interest-oriented content requirements. It was a preview of the 1996 Telecommunications Act, which further invited the concentration of media ownership. The changing media environment, ever more dominated by the rip and tear of the free market, would precipitate many other science journalism collapses in the coming years, making the early 1980s popular-science bust seem a sign of things to come.

Sagan was prescient in recognizing the great harm that Reagan's contempt for science would do to American society and in understanding the need to combat it. By all rights, he should have been a hero to the scientific establishment and a role model for younger researchers, all the more so because even as he succeeded, other science popularizers were struggling and failing.

Yet instead, Sagan was punished by the scientific community for his public endeavors. The persecution began as early as the 1960s, when Harvard University denied him tenure. Nobel laureate Harold Urey, a chemist who had previously served as one of Sagan's mentors, helped quash his chances with a nasty letter objecting to Sagan's budding media and outreach efforts. Already, young Sagan was getting lots of press attention. "Quite a few scientists in those days didn't feel it was right to have that sort of publicity," remarked Sagan ally Fred Whipple, who had tried to keep him at Harvard but failed.

History repeated itself in 1992, when Sagan was nominated for membership to the highly prestigious National Academy of Sciences. It was an honor requiring distinction in original scientific research, which Sagan could certainly claim—he had made significant contributions, for instance, to the study of the Venusian greenhouse effect and Martian dust storms. But Sagan's nomination proved divisive, and with only about half the academy reportedly voting in his favor, he was ultimately denied admission. According to reports from within the academy, Sagan was criticized for "oversimplification" in his scientific writings. Lynn Margulis, an academy member and by then Sagan's ex-wife, wrote him a sympathetic note in the wake of the debacle: "They are jealous of your communication skills, charm, good looks and outspoken attitude especially on nuclear winter." The scientist who nominated Sagan, the distinguished origins-of-life researcher Stanley Miller, concurred with her assessment. "I can just see them saying it: 'Here's this little punk with all this publicity and Johnny Carson. I'm a ten times better scientist than that punk!'"

In their treatment of Sagan, the nation's leading scientists had made clear their view of popularizers within their ranks, and of public outreach generally. It was a fateful position. The seeds Reagan had planted would bear copious fruit during the Bush administration, when science would come under aggressive attack and find too few ready troops lined up for its defense.

CHAPTER 4

Third Culture, or Nerd Culture?

THE 1990S WERE A DIZZYING, PARADOXICAL TIME FOR AMERICAN SCIENCE and those who cared about it. The Internet, born out of government-funded research, was creating opportunities like never before. Teenagers and twenty-somethings were getting fantastically rich from it, devising ingenious start-ups in their dorm rooms, while Bill Clinton was extolling it in his speeches.

Science was "hot" again, or so it appeared. The heyday of the science magazines and prime-time television shows may have passed, but science books seemed to be breaking through as never before. Sagan had virtually created the mass science-book market with *Cosmos,* and physicist Stephen Hawking did one better in 1988 with the publication of the mega-best-seller *A Brief History of Time*. Other science titles began making increasingly frequent appearances on best-seller lists, works by such authors as Richard Dawkins, Daniel C. Dennett, Jared Diamond, Richard Feynman, James Gleick, Stephen Jay Gould, Lawrence Krauss, Steven Pinker, and E. O. Wilson.

But while hard science was crowding bookshelves, paranormalist schlock was filling the airwaves, as the mass media sold the public psychics and UFO conspiracies, often by airing pseudo-documentaries that strategically blurred the line between fact and fiction. It went far beyond dramas like Fox's *The X-Files,* the dynamic-duo saga of a paranormal believer who always bests his scientific counterpart in their attempts to figure out what's *really* going on. None other than CNN's Larry King could be found regularly promoting spiritualist "mediums" such as John Edward and Sylvia Browne, and hyping the alleged Roswell conspiracy.

Even more worrying were developments on the political front. With the cold war over, government support for research in core fields like physics seemed newly vulnerable. In 1993, Congress killed funding for a much-desired "big science" project—the particle accelerator better known as the Superconducting Super Collider—amid admonitions that with the collapse of the Soviet Union, scientists couldn't expect to get all these big toys anymore.

At the same time, the Republican Party continued to define itself in adulation of Reagan and in opposition to mainstream science. After taking control of Congress in 1994, Newt Gingrich's Republicans proceeded to kill the legislative body's scientific advisory office and, with it safely disposed of, to challenge well-established scientific positions on subjects including the depletion of the ozone layer and the emerging problem of global warming. The newly empowered Republicans also proposed to slash research budgets dramatically and even to do away entirely with some government science agencies, for example, the U.S. Geological Survey. Meanwhile, a newly rejuvenated clique of anti-evolutionists, soon to become widely known as the "intelligent design" movement, began to stir.

In *The Demon-Haunted World* (1996), the final book published before his death, Carl Sagan worried openly that the forces of darkness were beating out those of scientific enlightenment:

> I have a foreboding of an America in my children's or grandchildren's time . . . when awesome technological powers are in the hands of a very few, and no one representing the public interest can even grasp the issues; when the people have lost the ability to set their own agendas or knowledgeably question those in authority; when, clutching our crystals and nervously consulting our horoscopes, our critical faculties in decline, unable to distinguish between what feels good and what's true, we slide, almost without noticing, back into superstition and darkness.

These are the words of a man who worried about far more than the lack of any obvious heir to his legacy. The warning signs were clear, but much of the scientific community failed to heed them—and instead spent the better part of the decade taking actions that would ultimately only diminish their status as public intellectuals.

The success of *A Brief History of Time* and other serious works of science helped fuel a period of triumphalism—even, arguably, hubris—among some members of the scientific community. In his 1998 book *Consilience: The Unity of Knowledge,* for example, the esteemed Harvard biologist E. O. Wilson asserted that science would ultimately provide the means of tying all the academic fields together; other disciplines would gradually accede to its tremendous explanatory powers. Such statements got a boost from the state of science itself. New insights in fields such as cognitive science, complexity theory, and evolutionary psychology seemed to promise sweeping new ways of reducing human nature and even culture to genes replicating, neurons firing.

In 1991, John Brockman, the innovative literary agent to the scientists, went so far as to proclaim (quoting Stewart Brand) that science provided the "only news" out there. Brockman further asserted that a group of scientists writing directly for the book-buying public—people like Wilson, Gould, Pinker, and Dawkins—composed a "third culture," a new clique of cutting-edge intermediaries between the academic intelligentsia and the general population. These scientists, Brockman argued, had surpassed experts in all other fields in their ability to explain the mysteries of human experience:

> The third culture consists of those scientists and other thinkers in the empirical world who, through their work and expository writing, are taking the place of the traditional intellectual in rendering visible the deeper meanings of our lives, redefining who and what we are.

The bold claims of the third culture czars sprang from genuine enthusiasm for (and excitement about) promising new areas of scientific research, but there was also an occasional undercurrent of arrogance and superiority that led the movement in less constructive directions. The third culture's frequent attacks on religious belief were perhaps most damaging. Although the creationist movement had certainly set up the false dichotomy between science and religion, a handful of influential public scientists welcomed and inflamed the battle.

Among third culturists, Richard Dawkins and Daniel C. Dennett in particular were fond of fighting faith, even as they advanced various defenses and explications of evolutionary science and sought to claim new intellectual territory for Darwin. In a 1991 essay, Dawkins likened religious belief to a "virus of the mind." In his 1995 book *Darwin's Dangerous Idea,* meanwhile, Dennett advanced the idea that the theory of evolution is a "universal acid"—it cuts through many cherished preconceptions, especially religious ones. Later, the two would also endorse a tin-eared campaign to relabel atheists like themselves "brights"—the obvious implication, of course, was that believers were dim (Dennett and Dawkins lamely protested that they hadn't meant to imply that, but the damage was done).

American culture at the time was moving in a very different direction. The Republican Party had been fully engulfed by the evangelical right, and Democrats were triangulating. By overwhelming majorities, Americans professed a belief in God. The third culturists were selling books, but bridging divides was another matter entirely.

In fact, throughout this period, public opinion surveys taken repeatedly, with invariant question wording, showed that an alarming percentage of Americans agree with this stunning statement: "God created human beings pretty much in their present form at one time within the last 10,000 years or so." Roughly 46 percent of the public holds this anti-evolutionist, young-Earth-creationist, and scientifically illiterate view. That number has held constant since 1982, the first year in which the question was asked, apparently untouched by the waxing and waning of popular-science efforts, whether through magazines or best-selling books.

And the third culturists and their sympathizers didn't merely take aim at the faithful. They trained some of their biggest guns on their colleagues across the quadrangle in humanities departments. Rather than countering the enemies Sagan had identified, they fired inward at other intellectuals.

During the 1990s, some humanists—particularly in the fields of literary criticism and cultural studies—had become heavily invested intellectually in "postmodernism," an incredibly broad term that literally means nothing more than "after modernism." By and large, when critics assailed postmodernist thinking, they were really attacking post-structuralism, a French philosophy pioneered by figures including Jacques Derrida, Roland Barthes, Michel Foucault, and Julia Kristeva. Emerging in the 1960s, post-structuralism was closely tied to radical politics; though the post-structuralists were united by no single credo or manifesto, theirs was a broad-based critique of bourgeois forms of authority—including science. The idea of objectivity, they claimed, was merely one of the means by which power was enacted. Assertions about "truth" or "reality" were socially constructed, inevitably the product of particular ideologies, social conditions, and power structures. Mythology, by this logic, is neither more nor less truthful than the laws of physics.

Precisely because science seeks to reveal objective truths about the way the world works—and claims that the laws of physics are verifiable in a way that the claims of mythology are not—post-structuralism was anathema to many scientists, who felt roused to defend their profession. The first major fusillade in the so-called Science Wars came in 1994 with a book by biologist Paul R. Gross and mathematician Norman Levitt entitled *Higher Superstition: The Academic Left and Its Quarrels with Science*, a full-scale attack on academic "po-mos" that didn't mind engaging in a little ridicule now and again. "Since much of what we write will appear, and in places may actually be, polemical, a certain gleefulness may be imputed to some of our observations," the authors warned, and then claimed they were criticizing "not enemies but friends." (So much for that friendship.) Gross and Levitt argued that the "academic left" had embraced "open hostility toward the *actual content* of scientific knowledge and toward the assumption, which one might have supposed universal among educated people, that scientific knowledge is reasonably reliable and rests on a sound methodology." But the postmodernists weren't launching incisive critiques, wrote Gross and Levitt; rather, they'd succumbed to "muddleheadedness." They didn't even understand the science they presumed to criticize.

Bizarrely, despite their book-length attack, Gross and Levitt admitted scientists didn't really have much to fear from the academic left. The threat of postmodernism was not so much to science per se, they claimed, as to the "larger culture," which would be somehow defiled if the academic left's esoteric critique became "the dominant mode of thinking about science on the part of nonscientists."

Higher Superstition takes a belittling tone toward the humanities in general, not just the small percentage of humanists engaged in post-structuralist critique. Gross and Levitt concluded their book by suggesting that if humanists were to leave campuses *en masse*, scientists could simply teach their courses for them; the reverse situation being, of course, inconceivable. They even went so far as to argue that many scholars who flirted with postmodernist modes of analysis and applied them to science were guilty of "intellectual dereliction" and to suggest that perhaps scientists themselves ought to be involved in the "hiring, firing, and promotion" of these people. The threat was at once insular and yet deeply aggressive; call it an ivory-tower war chant.

Higher Superstition was just the beginning. In 1994, the New York University physicist Alan Sokal, partly inspired by Gross and Levitt and aiming, as he later explained, to save his allies on the political left from what he viewed as an intoxication with disastrous nonsense, decided to take action. So he submitted a prank article, entitled "Transgressing the Boundaries: Towards a Transformative Hermeneutics of Quantum Gravity," to the left-wing academic journal *Social Text*. The article basically amounted to anti-science gibberish, but written in a cleverly, and at times hilariously, self-referential way. For instance:

It has thus become increasingly apparent that physical "reality," no less than social "reality," is at bottom a social and linguistic construct; that scientific "knowledge," far from being objective, reflects and encodes the dominant ideologies and power relations of the culture that produced it; that the truth claims of science are inherently theory-laden and self-referential; and consequently, that the discourse of the scientific community, for all its undeniable value, cannot assert a privileged epistemological status with respect to counter-hegemonic narratives emanating from dissident or marginalized communities. . . .

Although the editors of *Social Text* initially held the piece and asked for revisions—which, they say, Sokal refused—they ultimately published the article in 1996 in a special issue on the Science Wars. They soon came to regret their decision: Once the article was published, Sokal promptly exposed the hoax in *Lingua Franca* magazine, and *Social Text*—and indeed postmodernist critique in general—thereupon became a laughingstock. And a very public one at that: The story became national news, making it all the way to the front page of the *New York Times*.

And still, the Science Wars raged on. The next sally came in 1998, with the publication of E. O. Wilson's *Consilience*. In it, Wilson announced that "the greatest enterprise of the mind has always been and always will be the attempted linkage of the sciences and the humanities"—and then tried to pull it off. Although he had contributed a favorable blurb for Gross and Levitt's book, Wilson wrote calmly, respectfully, and indeed, beautifully; neither his tone nor his actions were a provocation. But his intellectual position certainly was. He argued that the social sciences and humanities could eventually reduce to scientific principles. Even the arts and literature, as products of human beings, were ultimately grounded in and therefore explainable in terms of biology, chemistry, and physics. All of human knowledge could thus eventually be unified based on scientific foundations; much of the social sciences would simply become biology, and the humanities and science would also "partly fuse."

Wilson's claims certainly represented a stunning scientific land grab, yet they are also highly questionable. Although neuroscience, for instance, has explained some aspects of brain functioning, it's less than clear that it will ultimately be able to reduce all of human thought—the imagination, the nature of memory—to mere patterns of neural firing. If anything, developments in the sciences have created greater need for humanists, especially in the realms of ethics, philosophy, and policy, rather than annihilating or subsuming them.

The great problem with the Science Wars wasn't that they were ineffective but that they were ultimately irrelevant. The influence of poststructuralism within the academic realm peaked in the 1990s and has been declining since—not because of Alan Sokal or *Higher Superstition*, but because that is the way academic trends work. Even Sokal himself acknowledges that the postmodernist threat—if there ever was one—passed very quickly. "Back in the 1990s," he has written more recently, "conservatives could still make rhetorical headway by insisting that postmodernist academics posed a dire threat to reason and scholarship. This was always an exaggeration . . . ten years on, that zeitgeist is unrecognizable." But all the energy spent fighting the Science Wars distracted from the real enemy at the gate—the dumbing down of American culture.

The 1996 Telecommunications Act passed with overwhelming bipartisan consensus (91–5 was the Senate vote). Promising lower costs to consumers thanks to more competition within the media industry, the bill generated little debate and almost no attention beyond the Beltway. It should have been seen as a disaster for American intellectual life, for serious and responsible public-interest journalism, and for science in particular.

The mass media had always been driven by a profit model, but with regulation by government to ensure that the public airwaves weren't fully flooded with lowest-common-denominator programming. Yet by deregulating media companies and relaxing limits on ownership, the 1996 law further advanced an ongoing trend of mergers and consolidation: the buying up of local radio stations by huge companies like Clear

Channel, and of television stations, cable networks, and other media branches (from film to publishing) by Viacom, Disney, Time Warner, News Corp, and others. These large conglomerates were beholden to shareholders who wanted to see profits grow, and they often opted to perform this nonpublic service by cutting costs or homogenizing content across their many branches. Simultaneously, it became much more difficult for the public to object to such trends, because the Telecommunications Act also lengthened broadcasters' government-granted license terms and made it harder for them to lose them.

How did such developments affect science? In the same way they affected the media's treatment of many serious, complex topics, such as international affairs, that naturally have a hard time competing with less substantive fare (like celebrity news and infotainment). As gigantic media companies eyed the bottom line, informative or educational science content would often be among the first things that didn't seem worth retaining. Meanwhile, the ongoing growth of cable television fractured audiences, luring many viewers away from substantive content even as the news that still remained took on a politicized tilt (e.g., Fox News) or devolved into partisan shout-fests. So even while scientists were squabbling with humanists within the academic sphere, the ability of either group to reach the broader public through the media was becoming an increasingly difficult proposition.

The media industry transformations of the 1990s and beyond represent just one of the ways in which the Science Wars would soon seem irrelevant. An even bigger shock came in the form of concerted and wide-ranging political attacks on science from the Bush administration, an assault that everyone suddenly recognized as a much bigger deal than anything the "academic left" had ever done. By 2004, E. O. Wilson could be found as a prominent signatory on a statement, organized by the Union of Concerned Scientists, denouncing the new administration for inappropriate political interference with scientific information across a broad array of hot-button issues, including global warming, sex education and contraception, endangered species protections, and much else. Science was now unequivocally under attack by those in power in Washington; in contrast, postmodernists had never even had enough power to run university campuses.

In this context, who should offer an olive branch but the disciples of the humanities? Times had been at least as hard for them as for science; in English departments, for instance, student numbers and academic resources were steadily dwindling. Still, they made time for a little solidarity. The summer 2005 issue of *The American Scholar*, a leading journal for humanists, blazoned the phrase "Science Matters" on its cover. Inside, editor Robert Wilson explained to readers that although "the attack on science has always been *our* game . . . the enemy of our enemy is most definitely not our friend." The right's attack on science, Wilson continued, "is an attack on reason, and it cannot be ignored, or excused, or allowed to go uncontested." Soon even Bruno Latour, a leading sociologist of science who had been critiqued by Gross and Levitt and by Sokal, could be found fretting over right-wing ideologues' attacks on the scientific consensus on global warming, and the way they sought, "artificially," to maintain a controversy where none actually existed within the scientific community.

In passing, Latour also made this revealing, and sad, remark: "It has been a long time, after all, since intellectuals have stopped being in the vanguard of things to come." But after considering the events discussed above, it's hard not to blame that decline in part on the scientists and intellectuals themselves. Although there are many interesting and legitimate debates to be had between different academic disciplines, the Science Wars ultimately reflected little more than the narcissism of petty differences. Scientists and humanists may have very different approaches in their pursuit of knowledge, but both groups are fundamentally dedicated to the idea that knowledge *matters*. Their real but largely unrecognized foes were, and remain, the forces of anti-intellectualism that lie just beyond the ivory tower—and it is long past time to face them, squarely and effectively, unified with allies across our society.

The fact is, the underlying dynamics affecting the cultural fortunes of science and intellect alike—particularly in today's media environment—are only growing worse.

PART 2

DIFFERENT RIFTS, STILL DIVIDED

Mike Roark (Tommy Lee Jones):
Find me a scientist! A geologist! Someone who
can tell me what the hell is going on!

—*VOLCANO*, 1997

CHAPTER 5

Science Escape 2008

OCTOBER 4, 2007, MARKED THE FIFTY-YEAR ANNIVERSARY OF THE SOVIET launch of Sputnik, the event that catalyzed the U.S. government to make a massive investment in science whose impact lasted for decades. It is profoundly disheartening to contemplate how far we have fallen since.

And so, hoping for nothing less than to reintegrate science into the broader public discourse and pull it back to the center of our culture, we went to work—roughly a month after the Sputnik anniversary and a year before the 2008 presidential election—on a grassroots initiative called ScienceDebate2008. In essence, it was a collective, nonpartisan call for the presidential candidates to publicly debate science and technology policy on the campaign trail, before a national television audience.

The idea for this push originally came from Matthew Chapman, a screenwriter, movie director, author, and science aficionado who also happens to be the great-great-grandson of Charles Darwin. It was spearheaded by a motley group of people throughout the country who organized out of common interest and motivation without any initial funding or institutional structure, and who were centrally led by another screenwriter and political strategist, Shawn Lawrence Otto. The initiative's message was splendidly direct: Science matters, to policy and to the economy. Therefore, politicians ought to debate science policy if they aspire to be president.

After eight years of enduring George W. Bush's hostility to science and amid growing fears that the United States could be falling behind in science and innovation, the idea of ScienceDebate2008 resonated deeply with the American scientific community. Before long, we had brought on board, as cosponsors, the leading institutions of American science—the American Association for the Advancement of Science and the National Academy of Sciences—as well as the umbrella Council on Competitiveness, which represents university presidents, labor leaders, and corporate CEOs. In addition, ScienceDebate2008 had garnered endorsements from scores of Nobel laureates and scientific luminaries, such as biomedical innovators David Baltimore and Harold Varmus, and physics pioneer (and now Obama energy secretary) Steven Chu; prominent political figures ranging from Newt Gingrich to Obama transition team head John Podesta; and 38,000 individual Americans. Virtually the entire community of American science rallied behind the cause, and with an extraordinary speed and passion that demonstrated how much this community, after the Bush years, had come to yearn for a higher profile in decision making.

Alas, when it came to science, the presidential race showed little hope of improving on the past. By early 2008, for example, TV's top five Sunday talk-show hosts—the late Tim Russert, George Stephanopoulos, Wolf Blitzer, Chris Wallace, and Bob Schieffer—had interviewed the various candidates more than 175 times and asked some 3,000 questions. Yet only six of those exchanges even mentioned "global warming" or "climate change." And the candidates didn't want to talk about science policy any more than the media cared to ask about it. Despite securing broadcast partners in PBS's *Now* and *Nova*, locking down a desirable venue—Philadelphia's illustrious Franklin Institute, named after the great scientist and founding father—and suggesting an ideal date of April 22, just before the Pennsylvania primary election, ScienceDebate2008 found its invitation declined by the Hillary Clinton and Barack Obama campaigns and ignored entirely by the John McCain campaign.

And just as ScienceDebate2008 couldn't get any traction with the campaigns, so the group also struggled with the political journalists who set the national agenda. ScienceDebate2008's formation, and its demand for a presidential debate on science policy, represented news by any reasonable standard: It was both unprecedented and highly policy relevant. The nation's brain trust wanted to hear from the politicians, and that's not the kind of invite John F. Kennedy or Dwight D. Eisenhower would have turned down. Yet although blogs, foreign media, and science journalists were fascinated by the story—*Le Monde* covered it, for instance, as did Radio New Zealand; and of course outlets like *Scientific American* and NPR's *Science Friday* dug in—the American mass media largely ignored it, presumably lumping the scientists together with all the other interest groups demanding attention from candidates.

Although Clinton and Obama were perfectly willing to attend a "compassion forum" to discuss "faith, values, and other current issues" at Pennsylvania's Messiah College during the primaries, and McCain and Obama both appeared at a "civil forum" at celebrity pastor Rick Warren's Saddleback Church during the general election, no science debate ever transpired. Obama and McCain did eventually respond in depth to fourteen written science policy questions posed by Science Debate2008, but there were no live speeches, no interviews, no chances for follow-up from the candidates, and no televised events to put science on the national agenda during one of the most closely followed elections in American history.

The ScienceDebate2008 experience should give pause to any scientist—or civilian—who feels that the election of Barack Obama will single-handedly solve the problem of America's scientific illiteracy. It's true that our current president evolved, over the course of the campaign, into a leader who shows a deep appreciation of science—in significant part, we believe, because he and his advisers were continually dogged by ScienceDebate2008—and has now begun to govern that way. Still, the problems encountered between scientists and politicians during the

2008 election were structural and systemic in nature, and not likely to disappear simply by virtue of any single candidate's victory.

Politicians, political strategists, and political journalists either didn't grasp the importance of talking about science in an electoral context or, worse, they feared it. ScienceDebate2008 CEO Shawn Otto says of his interactions with the campaigns: "[They] were terrified of [the debate idea]. They saw it as a high risk proposition, and maybe a sandbagging. Even if it wasn't, it would require lots of prep time and huge political exposure in order to move a relatively niche audience. If someone made a mistake, the thinking went, they would open themselves up to potentially fatal ridicule."

Scientists thought a presidential science debate was a great idea because they assumed that the rational airing of policies and differences should lead to better decision making and wiser voting, not to mention higher-profile treatment of critical science issues. Yet politicians viewed it as a lose-lose proposition; and since it clearly wasn't going to happen, political journalists viewed it as irrelevant—not a story.

Despite Barack Obama's pledge to restore science to its "rightful place" in Washington, then, scientists have a long way to go if they're to fully regain the political ground they've lost since October 4, 1957.

The late 1950s and early 1960s were, after all, a time when prevailing geopolitical circumstances sent the nation's leaders running to scientists for help and the scientists answered. As Sputnik faded from view, however, the politicians went back to being politicians. And the scientists, now the recipients of heavy federal funding from the taxpayer for their research, pursued a strategy of studied political detachment virtually unique in American public life: They would remain "on tap" to deliver their advice to politicians but aloof from direct electioneering. As longtime science journalist Daniel Greenberg puts it: "With very rare exceptions, [scientists] don't run for office or organize under their professional identities, as lawyers, physicians, bankers, and others regularly do."

Such a stance made perfect sense when it originated in the 1950s. But since then, as politicians have grown less needy and solicitous, it has fueled science's declining political influence. In the absence of a clear and urgent need for scientists and politicians to work together closely, the two groups have instead largely settled into operating as separate cultures with far too little understanding of one another or productive interchange.

In part, the divorce of science and politics can be explained by the very different worldviews that inform each field. Scientists look at the world and see order, and generally assume rational actions will (or should) be taken. They go to painstaking lengths to prove or justify their recommendations by quantifying and calculating possibilities, by modeling and accounting for as many inputs as possible within a system. They study universal principles and global problems. And as C. P. Snow observed fifty years ago, they have the "future in their bones": They take a very long-range outlook on where things are headed, even modeling what's to come through the haze of uncertainty. A classic example is the 100-year or longer projections now used by climate scientists to forecast the catastrophe we are courting if we don't do something—fast—about our greenhouse gas emissions.

Politicians live in a very different world, one in which they are more often rewarded for playing to voters' emotions rather than their intellects. Even if they themselves know better, they recognize that particularly in the television age, charisma, charm, and personal appeals will get them a lot further than logical argumentation. Unlike scientists, politicians respond to constituencies that are local—a city, a congressional district, at most a nation that covers only a small percentage of the world's landmass—rather than global. Scientists study forces with outcomes ten, 100, even 10,000 years hence, whereas politicians function in the relentless and unforgiving world of election cycles and the permanent campaign.

These differences alone would be sufficient to explain the current chasm between America's scientific and political communities. But there is also the problem of specialization. Whereas good science is rewarded for being painstaking and nuanced, politics is the enemy of subtlety—political battles are fought out in sound bites, decided in up or down votes. In this context, the politician often suspects that the scientist cannot see beyond his or her narrow specialty and spends too much time on minutia. And politicians find nothing more maddening than when scientists refuse to come out and say what they really think, hiding behind a veneer of "objectivity." Hence the recurrent congressional joke about the need to find a "one-handed scientist" to give testimony: a scientist who won't constantly say, "On the one hand . . . on the other hand."

This is a problem Sheril experienced personally as a fellow in the 109th Congress, where she observed that the groups arriving at her senator's office to make their case on science-related legislation seemed completely out of touch with the environment on Capitol Hill. Technical experts touted significance values and statistical figures that were at times so obscure that even Sheril couldn't understand the data, despite her experience in academia. Other science lobbyists promoted idealistic solutions without a grasp of the socioeconomics or people involved in the situation. Meanwhile, those on the "other side"—global warming "skeptics" and oil lobbyists, to name a few—were articulate and well organized, with a far better understanding of how to appeal to a congressional audience. They worked with each other, seeming to agree on goals, methods, and the take-home message. They were *effective*.

The fundamental problem was apparent: Too many scientists had internalized the idea that in some fundamental sense, they had to stay "above" politics, as though it were something dirty. Indeed, in the research community there's a commonly held belief that involvement in the legislative process tempts one to "become an advocate"—the enemy of objectivity. Political engagement, many scientists fear, can damage one's reputation. And of course it can also detract heavily from the time spent on research.

Many scientists also believe that political engagement is a waste of time, and unfortunately, they are often right. Politicians are notorious for their cynical use of scientific information; in the U.S. Congress, where only 8 percent of elected officials hold a science or medical Ph.D., scientific studies are regularly used as an excuse for doing nothing. Calling for "more research" is an excellent way of punting. At the same time, politicians are notorious for digging up scientific "facts" that appear to support what they already wanted to do anyway.

Such cherry-picking is easy to do, even inviting: Congress is awash in information, much of it questionable or self-interested, emanating from think tanks, advocacy groups, bloggers, journalists, lobbyists, and many others. With so many reports, agencies, and institutes on hand providing "expertise," information overload makes for a constant struggle, on the part of non-scientists, to weed through it all and determine what to trust. There's no official sifting mechanism, either: The previous one, the congressional Office of Technology Assessment (OTA), died at the partisan hands of the Gingrich Congress in 1995, and thus far hasn't been resurrected by the now-majority Democrats. And even OTA, it was often complained, worked too slowly most of the time to serve the congressional schedule—as do universities and the National Academy of Sciences.

In all of these ways, scientists and politicians regularly reenact the problem of the "two cultures." Sometimes the consequences of the disconnect are vast and egregious, such as the twenty-year failure to address climate change. And sometimes they're simply comical. For instance, Representative Vernon Ehlers (R–MI), one of three physicists in Congress, describes having to rush to the floor to prevent fellow members from killing science programs they haven't understood—assuming, for instance, that "game theory" research involves sports.

Nonetheless, the idea of game theory being debated, however ignorantly, on the floor of Congress is less frightening than a much more common problem: the massive difficulty of getting science on the political radar in the first place. According to a 2008 report from the Keystone Center, a Colorado-based policy institute that interviewed leading members of Congress and their staffs to assess how they thought about science, most members "seem to have little care about, interest in, or attention to technical and scientific matters in particular, and to legitimate and credible sources of information to guide Congress on such issues when it chooses to take them up." In fact, the Keystone report found that because many decisions about science policy are made not through decision or analysis but on ideological and political grounds, "even if a credible, centrist analysis is conducted on a particular issue for Congress, it might not matter much in the current way decisions are made." The report even quoted one congressional interviewee as follows: "No one in Congress senses the need for science in their daily lives."

It's important to remember that when it comes to their treatment of science, politicians reflect their constituents and the rest of society. They rarely denounce science outright, any more than average Americans do. Rather, they use it and abuse it as convenient, because too often, that's all science means to them: It's a tool to achieve an end. On the surface, science appears beloved; beneath it, hardly considered, save among those few legislators who work directly on setting science policy and funding levels.

And so despite indications that the relationship between Washington and the scientific community will greatly improve under the Obama administration, this is no reason for scientists—or the rest of us—to feel complacent. Scientists shouldn't stake their political future, or that of the country, on the vicissitudes of changing administrations. Having painfully witnessed their political vulnerability, they should work actively to reduce it and to reach out to politicians, rather than assuming they'll come around simply out of duty or interest.

The politicians *should* heed science, and care about it, without anyone asking or beseeching. But as we don't live in a perfect world, scientists must also strive to make their knowledge relevant, something that hardly happens automatically. It takes a lot of work and experience to shape scientific information in a politically useful way; and in general, the scientific community has not invested much energy in creating specialists capable of carrying out this culture-crossing endeavor.

Small efforts can go a long way, however—after all, effective communication isn't rocket science (or neurobiology, or particle physics). Thrust scientists into political contexts often enough, and they'll pick up a great deal. Programs like the American Association for the Advancement of Science's congressional fellowships or the John A. Knauss Sea Grant fellowship program do precisely that by inserting scientists into congressional offices. But what they bring about informally should be done far more systematically, as a central priority of the scientific community. The goal should be nothing less than to redefine the role of the scientist in public affairs. This requires a shift away from the identity of a lobbyist or special interest who simply wants more research funding and toward that of a trained communicator and options broker who provides politically feasible ideas and can help politicians *succeed* through the use of science, and recognize why they need it.

In other words, as Daniel Yankelovich has suggested, we need a new caste of savvy scientists who can act as "framers" of policy issues. These scientists would understand the varied socioeconomic and political pressures that impinge upon the legislative process and know how to integrate accurate scientific information with a range of achievable and realistic policy options to facilitate the process of decision making. Rather than testifying before Congress or briefing legislators from a "just the facts" perspective, these scientists would be ready to address the questions most relevant to political leaders, questions involving American competitiveness, employment, innovation, and national security.

To succeed, such scientists will have to understand how to communicate effectively with politicians. To this end, they might take a page from Canadian politician Preston Manning, who has written incisively about the science-politics divide, differentiating between "source-oriented communicators" and "receiver-oriented communicators." Most scientists are source-oriented communicators: that is, they communicate in whatever way feels most comfortable to them, irrespective of the audience and usually in scientist-speak. Receiver-oriented communicators, in contrast, think about an audience and how to reach it, and only then determine the appropriate message to use. For scientists to play an effective role in American politics, they will need representatives skilled in this latter mode of interaction.

What are some "receiver-oriented" communication strategies that will help scientists in the political arena? Scientists, Manning argues, need to "establish a relationship with the political community on grounds other than the milk cow–milking machine relationship." In other words, they shouldn't merely reach out to politicians when they want research funding. Rather, they should establish long-term personal relationships that are multidirectional in nature, so that they are helping as much as being helped.

Second, in a world dominated by the twenty-four-hour news cycle, in which elections are won and lost not through personal outreach but by media appearances, scientists will have to find ways of presenting science issues in such a way that politicians will instantly recognize their media communicability. Scientists will have to accept that their advice is being judged not on its substantive content—at least not at first—but explicitly on the utility of its packaging. Politicians, Manning notes, are always thinking, "If we adopt that position, how will I explain that to the television reporter who is waiting outside this room when she sticks her mike and her camera in my face?"

Carl Sagan is the perfect example of a scientist who knew how to do this; indeed, he did it so well that he once contributed scientific language to President Jimmy Carter's farewell address. When needed, Sagan could put on his scientist-speechwriter hat. There ought to be many more scientists with this ability, and far more recognition in the scientific community that having such a valuable skill is in no way inferior to excelling as a researcher. The best way to ensure such recognition, of course, would be to explicitly count successful political outreach as part of a scientist's credits for job advancement—in other words, it should be favorably weighed in university tenure review. Meanwhile, young scientists in training should receive courses in public speaking and become acculturated to the idea that this is part of their job description.

Scientists who have many political encounters may develop such abilities independently, but simply assuming they'll do so on their own leaves far too much to chance. Instead, these skills should be systematically cultivated among a group of effective science communicators who can do this critically important work on behalf of their community.

And even as initiatives of this kind would go a long way in preparing scientists for public roles, a little innovation and experimentation can also achieve wonders. The experience of ScienceDebate2008, while disappointing in its failure to secure a national science debate, nonetheless reveals a great deal about how much can be accomplished with imagination and hard work.

The organization had almost no funding or staff. It was literally run by two screenwriters who needed something to do during the 2007–2008 Hollywood writers' strike. Yet precisely this fact—that the organizers hadn't been through the rigors of the scientific training process, didn't have to guard their objectivity all the time, could politically strategize and innovate, and were willing to try off-the-wall approaches—probably made them the most effective emissaries to the political arena that science could have. As the organization's CEO Shawn Otto (who, when not organizing scientists, writes films such as *House of Sand and Fog*) puts it: "It may be telling that it was two screenwriters—mass communicators—and not scientists themselves who got behind this initiative to get it rolling. We, as ordinary citizens, saw the problem very clearly; as mass market writers we knew how to engage the public on it, and because of the writers' strike, we had time to take a flier and try to do something about it."

Consider Otto's actions after he was initially rebuffed by the campaigns in his request for a presidential science debate. He didn't give up: Rather, he set about proving to the candidates that they needed to participate, and that it would be good for them as well as for the country.

The push involved two strategies. First, working with the biomedical science advocacy group ResearchAmerica, ScienceDebate2008 under Otto's leadership commissioned a poll whose results suggested that science issues aren't necessarily a "niche" area after all. Eighty-five percent of the public said it supported the idea of a presidential science debate. Although we shouldn't make too much of this number, it certainly gave the presidential candidates something to consider, and in a language they understand—that of polling. If those results were correct, then maybe, just maybe, they wouldn't be written off as nerds for participating in a science forum.

Meanwhile, ScienceDebate2008 drafted, with the help of its members, fourteen top science policy questions that the candidates ought to answer, such as, "What policies will you support to ensure that America remains the world leader in innovation?" and "What is your position on the following measures that have been proposed to address global climate change: a cap-and-trade system, a carbon tax, increased fuel-economy standards, or research?" Putting such substantive questions down in print helped to prove that the proposed debate would not be some sort of pop quiz designed to embarrass the candidates if they couldn't name all the different kinds of quarks. Both steps were, essentially, examples of "receiver-oriented communication," assuaging the politicians' concerns about participating in a science debate and showing how that participation could be good for them, not just good for science in isolation.

Otto kept on pushing, and during the general election season, something finally clicked. As he relates:

> I made a last-ditch pitch to both of the campaigns saying, "Look, this is not a niche debate and we have the national polling data to show it. The questions are all out there and no matter who's elected these are several of the key problems you're going to be facing, and the voters have a right—and you have a responsibility—to assess your positions on these questions. You've got to at least answer them online, and ideally also in a televised forum."

The Obama campaign broke the ice: On August 30, 2008, it answered the fourteen questions in considerable depth, thanks to the help of a star-studded science policy advisory team the campaign had put in place by then. The Obama answers rippled around the Internet, placing pressure on the McCain campaign to do likewise. Two weeks later it followed suit, and then for the first time a science-centered point-counterpoint between the two candidates existed, and citizens could see that Obama's plans for addressing climate change involved a much more prominent international treaty component and that he wanted to double federal research budgets, whereas McCain looked more to the private sector to spur innovation. Soon the questions and answers were being quoted and referenced widely; ScienceDebate2008 estimates that they garnered attention in most major U.S. newspapers and in media around the world. ScienceDebate2008 didn't get the televised presidential science debate that it wanted, but in less than a year's time, thanks in significant part to Otto's work, it came out of nowhere and injected science policy into the campaign in a major, unforgettable way.

There are many lessons here—and many other forms of engagement with politics that ought to be tried beyond the ScienceDebate push, which will now continue in 2012. Those include getting more scientists to run for public office, perhaps even forming a science political action committee, or PAC, to support their endeavors. The field is wide open: Much remains to be tried.

But for now, let's give the last word to Otto, the screenwriter who, along with fellow writer Matthew Chapman (whose credits include *Runaway Jury*), did the most to make it all happen. Otto told us the story of what happened when he and Chapman ventured to Washington, DC, to rally supporters for the cause. They visited the Rayburn Office Building on Capitol Hill, meeting with House Science Committee chairman Bart Gordon (D–TN) and Minnesota Democrat Collin Peterson's staff, and as Otto reported, "They loved it." Then they left to catch a cab to talk to the National Academy of Sciences, and again, "They loved it." Maybe scientists and politicians aren't so hard to bring together after all.

At this point, it so happened that Chris called Chapman's cell phone, and upon learning what he and Otto were up to, exclaimed: "What a story—two screenwriters going to beg the National Academies to support a public debate on science!" Everyone had a laugh. And then, Otto recalls, "We hang up and I remark how it is a little bit like *Mr. Smith Goes to Washington.* The cabbie, who has on this big blue turban, turns around and says, 'I loved that movie!'"

CHAPTER 6

Unpopular Science

RICK WEISS WAS PROBABLY AT THE PEAK OF HIS CAREER. AMONG SCIENCE journalists filing stories at the national level, his byline ranked among the most recognizable, particularly on matters of biomedicine and bioethics. If you read a story about embryonic stem cell research in the *Washington Post* during the 2000s, it's likely Weiss penned it. For fifteen years, he served at the hallowed DC paper, a top outlet for science-centered journalism—or at least, it used to be. But it lost Weiss in 2008, when it did away with its dedicated science page.

Once a week on Mondays, the *Post* had long given science a full page, divided up into a full-length feature story and a short notebook section, with the remaining one-fourth devoted to advertising. In early 2008, however, that space got taken away. First it was cut to just half a page, then scaled back to every other week. "We fought in the science pod very hard against that change," says Weiss, who left the paper following the cut of the section, in part because of what he viewed as its declining commitment to science. "I took my arguments all the way up to the executive editor, but to no avail."

For Weiss, it was the culmination of a long and dismaying trend. He had already seen the number of *Post* science reporters decline and increasingly, the paper's remaining science reporters being pulled off the beat to cover something else. So along with 100 other *Post* reporters, he took a buyout and departed in mid-2008. "Science reporting at the

Post was obviously of shrinking importance to the editors," says Weiss. "I wanted to do more and better, and they wanted to do less. It's a sad symbol of how things are going."

In March 2009, as this book went to press, the *Post* science page reappeared, at least temporarily, apparently thanks in part to a fortuitous advertising arrangement. But Weiss was still gone, and the overall trend remains dismal. Indeed, almost simultaneously, the *Boston Globe* got rid of its own special Monday Health/Science section as a cost-cutting measure, a particularly stunning move in a city that leads the biotech industry. The trend for newspaper science journalism is definitely in the "less" direction. Papers today are hemorrhaging staff writers and slashing coverage areas as their business model collapses in the face of declining readership and advertising revenues, a consequence of the Internet upending the media landscape. In late 2008, the Pew Research Center for People and the Press released the finding everyone in the media world had been waiting for: "The Internet . . . has now surpassed all other media except television as an outlet for national and international news." For the young, the Internet now even rivals TV as the top news source.

As a consequence, many kinds of important newspaper coverage are disappearing: Foreign affairs reporting is struggling; many papers' previously respected Washington, DC, bureaus are vanishing. As for books, they're out, too, and in early 2009 the *Post* cut its stand-alone book review section. Meanwhile, those reporters who stick around the newsroom are increasingly required to become part-time bloggers, posting their stories online and then updating them throughout the day, expending their energy struggling to keep up with the twenty-four-hour news cycle rather than being able to dig in, investigate, learn, reveal.

Such trends have pummeled science reporting in particular, and though the *Post* and *Globe* may be among the biggest casualties so far, they're hardly the first. From 1989 to 2005, the number of U.S. newspapers featuring weekly science-related sections shrank by nearly two-thirds—from ninety-five to thirty-four. And many of the remaining sections shifted to softer health, fitness, and "news you can use" coverage, reflecting an apparent judgment by newspaper executives and editors that deeper and more thorough science coverage, or science policy coverage, simply doesn't support itself economically. The top "science" story found in newspapers these days is exercise and fitness; in one study, it accounted for 28 percent of total science-related stories.

And the problem isn't merely newspapers. Some specialized science programs notwithstanding, television is probably worse. Just one minute out of every 300 on cable news is devoted to science and technology, or one-third of 1 percent of coverage. Even as the *Post* and *Globe* were upending their science pages, CNN cut its entire science, space, and technology unit.

Surveying all this, you might think science has become less newsworthy or less relevant, yet nothing could be further from the case. We live at a time of pathbreaking advances in biotechnology and nanotechnology, of private spaceflight and personalized medicine, amid a global climate crisis, in a world made more dangerous by biological and nuclear terror threats and global pandemics. At the same time, advances in neuroscience are calling into question who we are, whether our identities and thought processes can reduce to purely physical phenomena, and whether we actually have free will. The media ought to be bursting with this stuff. Yet instead, career prospects for science reporters, the journalistic breed that cut its teeth covering the space race, have probably never been worse. As of 2005, only about 7 percent of the 2,126 members of the National Association of Science Writers had full-time positions at media outlets that reach broad publics—newspapers, popular magazines, radio, and television. The rest were freelancers, more specialized journalists, or public affairs officers for universities and other institutions.

All of this should more than suffice to show that there's a crisis today in the realm of science communication. What was always a difficult endeavor is only growing more so as market forces continue to dismantle public-interest-oriented, informative journalism of all types and supplant it with entertainment, blogging, or nothing at all. It's probably the single biggest threat to the role of science in American culture today—and it's one that scientists, and people who care about scientific thinking, have barely begun to combat.

Even before the current near-collapse of the newspaper industry, scientists and journalists had problems connecting. Much like scientists and politicians, the two camps have long operated as "two cultures" that rely on one another and yet start out from vastly different assumptions, making their interactions fraught, perilous, or worse. It's a dangerous schism indeed: After Americans leave the educational system, the media become their chief source of information about science. That's especially so for television media, the primary source of science content for 39 percent of Americans, according to the National Science Foundation. But the media's influence stretches much further. Top news outlets like the *New York Times,* the *Washington Post,* the *Wall Street Journal,* and the Sunday-morning talk shows set the broader media agenda, meaning that their treatment of science determines the responsiveness of the political system to it. In this context, lack of media attention signals a lack of priority—precisely what science has received. The prestige press's ignoring of the unprecedented ScienceDebate2008 initiative is actually quite typical. A study by the Project for Excellence in Journalism found that over a three-month period in late 2007 and early 2008 at the *New York Times* and *Wall Street Journal,* only about 1 percent of front-page stories were about science and technology (precisely 0.8 percent for the *Times* and 1.1 percent for the *Journal*).

In light of such findings, it's no surprise that scientists disapprove of how the media are covering their work. In a 2004 survey of the members of the Union of Concerned Scientists, for instance, 90 percent opined that the media did a "poor job covering science," and half reported having "difficult or disappointing experiences with the press." On the media side, sentiments can be similar: Some political or general-interest journalists even appear to feel disdainful of scientists, whom they regard as simply out of touch and incapable of communicating. An apparent throwaway line that appeared in a July 2008 *Newsweek* article about Charles Darwin and Abraham Lincoln (asking who was "more important," a typical journalistic conceit) says it all. "At least at the outset," wrote journalist Malcolm Jones, Darwin "was hardly even a scientist in the sense that we understand the term—a highly trained specialist whose professional vocabulary is so arcane that he or she can talk only to other scientists."

But the science-media divide springs from causes far more diverse than the mere fact that two groups don't always admire each other. Although they're broadly united by a search for the "truth," the practical ways in which scientists and journalists pursue that end vary widely. Media coverage tends to be episodic and event-driven, always in search of the dramatic and the new. In contrast, science is an ongoing process, at times exciting but mostly dull and monotonous, frequently characterized by false starts and dead ends and yet cumulative in its development of increased understanding. But journalists rarely write about or even perceive the process: They don't have time for covering the slow growth of knowledge. "In the newsrooms I know," writes *New York Times* science journalist Andrew Revkin, "the adjective *incremental* in a story is certain death for any front-page prospects, yet it is the defining characteristic of most environmental research."

Therefore, instead of relating the broad story of science in development, with its nuances and its policy implications fully limned, journalists more often pounce on some "hot" result, even if it contradicts the last hot result or is soon overturned by a subsequent study. As a result, members of the public can experience media "whiplash": The press tells them science says one thing, then it tells them science says the opposite. Should you drink more coffee, or less? Does global warming increase the number and intensity of hurricanes, or not? On these and many other topics, media reports have provided contradictory answers—both seemingly based on "science," of course—and many citizens have, as a result, questioned the reliability of journalists and scientists alike.

Another problem is that journalists hew to a relatively small number of often preconceived story lines—so-called angles or frames. There's the conflict story, the consensus story, the lone rebel going against the grain story: Journalists tend to approach any topic looking for one of these familiar narratives, and then filter all their further investigations and questions through it. That couldn't be more different from the scientific mindset, which is fundamentally oriented toward rooting out preconceptions. The "inductive" ideal of science, dating back all the way to the seventeenth-century English philosopher Francis Bacon, is to survey all the data (as much as is possible, anyway) before coming up with any "theory" to organize or explain them.

And there are still other sources of disconnect. If there's one thing that truly enrages scientists about journalistic coverage of the subjects they know intimately, it's the notion of "balance": the idea that reporters must give roughly equal time or space to the two different "sides" of a controversy. Especially prevalent in political journalism today, this media norm originally developed to help foster the semblance of objectivity, and certainly has its uses. But it can also serve as a crutch for reporters who can't devise more imaginative ways of structuring their stories. Balance can be misleading, even downright biased. In the media, granting attention implicitly bestows credibility, so should journalists really allot equal attention to the comments of the small band of scientists who deny the causal relationship between HIV and AIDS when the vast majority of mainstream researchers accept the connection between the two? Should they split column space between the perspectives of the few remaining global warming "skeptics" and the arguments of expert scientific bodies that have reached a consensus that it is caused by human activities?

Whenever journalists engage in such practices, you can bet scientists will tear their hair out. There's simply no equivalent, in science, to the journalistic norm of balance. Rather, scientific ideas advance on the basis of merit within a refereed publication process that demands rigor, accountability, and the ability to persuade colleagues that one's position best fits the available evidence. Thus, once scientists have become strongly convinced of something—that, say, the HIV virus causes AIDS—they feel insulted when journalists reopen the question outside of scientific channels.

But if journalistic norms really don't mesh well with scientific procedures, scientists' expectations of journalism can be unrealistic or even absurd. Consider a 2001 survey of 744 Dutch scientists, which found that nearly 80 percent believed journalists should comprehensively present the results of scientific research, though media time and space constraints make this impossible. In addition, more than 90 percent of the scientists felt that journalists should allow them to check their articles for accuracy prior to publication, a condition many journalists would refuse to submit to, as it calls into question their neutrality and objectivity. Indeed, half of the scientists further thought journalists should be *obliged* to make any changes that the scientist required to their story after reviewing it. No self-respecting journalist would agree to this.

Such mismatches help us understand how the media bungled the most important science-related story of our time: global warming. We were warned and warned about it, yet for decades did nothing while the problem steadily worsened. In large part, that's because the U.S. public continues to rate global warming as a low priority compared with many other issues and the politicians respond to that public—and both are getting their cues about what matters from the media.

Those media, in turn, have gotten the story wrong in multiple ways: first, by covering it as a "he said, she said" controversy during the 1990s (bowing to pressure from special interests and their pet scientists, who strategically attacked the scientific consensus) and then, even after moving away from such coverage, by providing far too little attention to the story overall, hardly proportionate to the grave dangers it poses. For global warming has the misfortune of being an "incremental" story. It keeps worsening, yes, but how often is it actually *news* of the kind that can outcompete all the other urgent matters demanding media attention?

The answer is rarely, if ever. In fact, although the level of attention devoted to global warming in the worldwide newspaper media rose sharply in 2005 and 2006, it has since fallen again into a decline, apparently triggered by the economic collapse, which has sucked much air out of other kinds of coverage. Scientists are growing increasingly terrified of what global warming could ultimately do—among other things, submerge our coastal cities—and are now contemplating the need for further meddling with the climate system (so-called geoengineering) as a last-ditch effort to reverse it. The press is AWOL.

Such cultural and professional rifts between scientists and journalists represent a long-standing concern, but the most profound problem for science journalism today is the full-scale transformation of the content industry that has occurred over the last decade. Not only are there fewer science journalists in the traditional news media; this fact itself means that the people still covering science are less likely to be specialized and well informed about it. This makes them all the more prone to the many pitfalls described above.

Although many scientists love to blast the press, today's writers and reporters probably deserve their sympathy more than their scorn. Even if they do care about science and want to do a good job covering it, they face unprecedented pressure to produce content that commands attention and ultimately helps sell the advertisements that keep their publications, stations, or programs afloat.

Scientists have traditionally been very little attuned to the business side of the media and how it affects them; they tend to view the press as having a high moral "responsibility" to cover research—period. In some sense, they think we're still in the age of Edward R. Murrow. In fact, it's the age of Bill O'Reilly.

It's not just the slow death of newspapers: The media industry has undergone epochal changes since the days when science enjoyed its cultural heyday in the United States. And these changes mean scientists can no longer assume that a responsible, high-minded media will treat their ideas with the decorum and seriousness they deserve, delivering them up to policy makers and the public for sober consideration. Instead, partisan media will convey diametrically opposed versions of where science actually stands on any contentious subject—consider the difference between how Fox News and NPR cover global warming—

even as most of the public (and many policy makers) will tune out science more or less completely in favor of other media options.

Underlying these developments are broad market-based trends in the media industry, themselves the result of two central factors: deregulation and—ultimately perhaps far more powerful—technological change.

Let's consider deregulation first. The trend that began during the Reagan administration, and continued throughout the Clinton years, shows few signs of abating. The mass media are dominated by a relatively small number of corporations, which have pulled together previously separate media sectors—movies, television, book publishing, music, magazines, radio, and many newspapers—and crammed them into their massive firms. Among the biggest behemoths: Disney (which owns ABC, ESPN, numerous film studios, and many other stations and entities), Sumner Redstone's CBS Corporation and Viacom (CBS, MTV, BET, Comedy Central, Paramount Pictures, and much else), General Electric (NBC Universal), News Corp (the *New York Post* and *Wall Street Journal*, Fox Television, Twentieth Century Fox, Fox News, the Fox Channel, HarperCollins Publishing, and that's just a small sampling), and Time Warner (CNN, *Time*, HBO, Warner Bros., and many other properties and brands). And alongside such "conglomeration" we've also seen, relatedly, "consolidation" or the concentration of media ownership: Once diversely owned local radio stations, television stations, and newspapers across the country are now held by only a handful of companies.

Deregulation and decades of mergers have brought us to this point, and a fundamental theme of these mergers has been to please investors. That means squeezing each individual station or newsroom in order to obtain the most profitable product. It often involves cutting back on staff, and cutting down on substance and quality as well—the homogenization of content, programming to the least common denominator. We're talking about the journalistic equivalent of replacing high-quality scripted television dramas, which require skilled actors and writers to produce, with endless cheap reality TV shows (something the con- glomerates are also doing).

In this climate of media conglomeration and consolidation, serious science coverage is often among the first things to go. Producing it re- quires seasoned journalists with high levels of training, who expect to receive salaries commensurate with their experience and expertise. These journalists are expensive, and media owners seem to keep mak- ing the judgment that they're not worth it, that science journalism doesn't draw enough eyeballs or, ultimately, advertisers.

And even as science coverage has been squeezed in the interest of improving the bottom line, we've also seen another trend: fragmenta- tion, or the loss of common media sources shared by large portions of the populace. This latter phenomenon is the broad consequence of technological change, which has certainly benefited media consumers by exponentially increasing their news and entertainment choices, and yet has also further impaired the translation of science through the media to the public and political leaders.

As recently as a few decades ago, there were a few authentic "news- papers of record"—the *New York Times*, the *Washington Post*—and they had not yet been besieged by bloggers and webzines. News on televi- sion, meanwhile, was confined to the broadcast networks, ABC, CBS, and NBC, and also to PBS. During television's so-called golden age, such outlets provided not only a shared cultural experience through a relatively limited number of entertainment programs, but also a shared news environment, including shared science news. Carl Sagan's *Cosmos* was a product of this era, and during the 1980s, retired CBS News an- chor Walter Cronkite even presided for two years over his own science show, *Walter Cronkite's Universe*.

There was no way it could last. First cable came along, providing hundreds of channel alternatives for those who wanted to disengage from serious news, and increasingly politicized platforms for those who remained plugged in—for example, Fox News and MSNBC. And it was just the beginning. The Internet's transformative powers may ulti- mately make cable's impact on the media seem trivial. Now newspapers are on the verge of extinction, but we have millions of blogs to suit every interest and political persuasion, and Google News to sift our headlines. If cable revealed a new continent of media choice, the Internet blasts us into a new galaxy.

The consequences are profound. Citizens who don't care much for science can easily escape it now. Ditto for citizens who are bored by politics: They can steer away from that particular channel or that particular corner of the Internet. The Food Network is waiting.

Among this smorgasbord of choices, science media outlets and programs like the Discovery Channel and PBS's *Nova* also exist, but only as one niche among many. Even that pinnacle of newspaper science journalism, the *New York Times*'s Tuesday science section—which as of 2006 employed over a dozen staff science writers and half a dozen science editors—only reaches perhaps a million people once a week, still a very small slice of America. Arguably, the most important news-oriented science communication today occurs via Comedy Central's *The Daily Show with Jon Stewart* and *The Colbert Report*, popular public-affairs-slash-comedy programs that manage to integrate a surprising amount of scientific content and treat it very sympathetically overall—as long as the scientists who go on air can laugh at themselves, and their profession, a little. Science can still be entertaining, at least at the hands of the very talented. But whether science-centered entertainment can substitute for science news and reporting is very much in doubt.

Although a combination of journalistic habits and media trends have considerably damaged the prospects for the widespread communication of science, it's not as though scientists remain blameless in this equation. Too often, instead of striving to adapt to changes in the media, they've opted to take themselves out of the game. It all goes back to the Carl Sagan phenomenon: The greatest science communicator in a generation was persecuted during his lifetime by much of the scientific establishment for being too much of a "popularizer."

Having interacted with numerous scientists, young and old, in the course of our work, we doubt many today would cheer the slighting of

Sagan. If anything, the episode is now seen as a skeleton in the closet for the scientific elite. Especially among the youngest generation of researchers—today's graduate students, recent Ph.D.s, and postdocs—we have seen a real hunger for training in media outreach. These scientists want to obtain the skills that can help them explain their work to a broader public; such abilities can only help them in a difficult and uncertain job market.

Yet even though the spurning of Sagan may not reflect the values of many modern scientists, that doesn't make them adequately prepared to communicate with and through today's massively challenging media. There remain at least two problems: first, a lingering anti-popularization sentiment among many more traditionally minded scientists, and second, institutional structures that fail to award successes in communication and thus create little incentive for scientists to engage in it.

The anti-popularization sentiment is, ironically, very much apparent in the burgeoning science blogosphere. When it comes to science and the media, the typical blog mode is to find an individual piece of science reporting with some particular failing and blast it—without addressing or even raising the broader issue of what's really going on in the media industry. One science blogger, University of Toronto biochemist Larry Moran, even put it this way in a discussion of the widespread job losses facing science reporters: "Science journalists have let us down. I say good riddance."

Not all scientists take such a lamentable line on science communication or ignore the changing media context in which it is now occurring. Many would like nothing better than to find a way to disseminate their knowledge to the broadest possible spectrum of the public. But even less ideological researchers face a practical problem. On an institutional and professional level, communication hasn't been much of a priority for scientists. Thus, we find them little trained in it, and little rewarded for engaging in it.

Herein lies the great difficulty, and the great opportunity. Universities must take the lead in institutionalizing new priorities within the world of science, so that communication becomes a central focus, not to the detriment of research but rather as a complement to it. Here, funding is essential. Institutionalization requires money, and that's why it's so unfortunate that an innovative piece of legislation introduced in 2007 by Representative Doris O. Matsui (D–CA) did not fully succeed. Matsui's legislation, originally the "Scientific Communications Act of 2007" and later made part of the America COMPETES Act, would have directed the National Science Foundation to start making grants to help graduate students in science obtain training in communication. But in its final form, the legislation wound up being pared down to a mere "Sense of Congress" without any dollars attached to it.

Just think: If a future bill puts some funding behind this idea, universities would start vying for it and setting in place new communication classes for scientists. Someone would have to teach them, meaning that someone would have to be good at it, so science departments would have to hire science-communication specialists or partner with communications and journalism departments. Either way, it's possible that before long, one significant criterion for career advancement in science would involve demonstrated success in science communication and in teaching about it. Perhaps professors could cite such successes in making the case for why they deserve tenure.

If the teaching of science communication were to win federal funding, and being good at it were to lead to promotion, things in science could change in a hurry.

There are other ways of addressing the crisis in science communication as well. One would be to circumvent market forces altogether by creating many more not-for-profit sources of science journalism and commentary, funded by tax-deductible, charitable donations, and meant to last for long periods of time, safely insulated from any type of market upheaval.

After leaving the *Post*, Rick Weiss went in that direction: He took a position with *Science Progress*, an online science policy publication of the nonprofit Center for American Progress. Chris writes there, too. It's part of a growing trend: For example, two hallowed and heavily endowed nonprofit educational institutions have each recently launched climate change and environmental journalism publications—Princeton has introduced Climate Central, and Yale now offers the online publication *Yale Environment 360*.

Such initiatives reflect their creators' realization that we probably can't hope for vast changes in the profit-driven media. And as we'll argue later, neither can we assume that science blogging, which is spreading rapidly and has many benefits, can fully replace what's being lost. In light of this reality, there's little other option for scientists and scientific institutions dissatisfied with the current state of affairs (and how could they not be?) than to take matters into their own hands.

The situation, in the end, is much like the problem science faces in the political arena. In both cases, scientists are the ones who know best what the other camp (politicians, journalists) is missing. Therefore, rather than assuming that today's media will dutifully carry information about science to the entire American public, it falls to scientists and their supporters to shift gears and carry their knowledge and message to the entire media.

And if that's still not enough, they'll simply have to create media of their own.

CHAPTER 7

Hollywood and the Mad Scientists

"REALITY ENDS HERE." IT'S THE UNOFFICIAL MOTTO OF THE UNIVERSITY OF Southern California School of Cinematic Arts, cast in concrete at the entranceway to the Robert Zemeckis Center for Digital Arts and, in Latin, at the South entryway of a new complex building. As scientist-turned-filmmaker and USC film school graduate Randy Olson explained to us, the slogan

is not a joke. It's a bold, challenging statement—a basic "screw you" to the outside world who thinks that accuracy and reality are important variables in storytelling, when in fact they are the dread enemy of storytelling. No storyteller wants an expert around who will question his premises. It's hard enough to tell a story without having some annoying expert sitting there correcting every detail and negating every premise.

Translated for more literal-minded scientists—who, when they think of *Star Wars,* think of the impossibility of having fire and loud explosions in space—the motto might be better rendered: Abandon All Accuracy, Ye Who Enter Here.

No wonder scientists haven't always been pleased with depictions of themselves, and the subjects they study, in the entertainment media. There's a long litany of complaints: Too many stereotypically nerdy scientists—think Rick Moranis in *Honey, I Shrunk the Kids*—and too few positive role models. Too many mad scientists trying to play God. Too many plotlines dependent upon the supernatural and the paranormal— stories in which, as in *The X-Files*, the credulous believer is always right and the scientist-skeptic is always wrong. And too many simply ridiculous "scientific" premises, epitomized by the 2003 film *The Core*, in which Earth's magnetic field begins to collapse because the planet's core stops spinning.

In light of the massive influence of the American entertainment industry, both here and around the world, scientists' complaints about Hollywood deserve to be taken seriously. Successful blockbusters reach audiences numbering in the tens of millions or more, and gross in the hundreds of millions of dollars at the box office. Some become pop-culture reference points for entire generations: Even those countless Americans who can't name a living scientist probably know who "Doc" from *Back to the Future* is. Entertainment industry expert Marty Kaplan, director of the Norman Lear Center at the University of Southern California's Annenberg School for Communication, perhaps puts it best when he describes Hollywood films as the "unofficial curriculum of society."

Within this curriculum, science usually gets taught through the medium of science fiction and the sometimes related disaster-movie genre, both of which have become dominant storytelling modes. From the *Star Wars* films to *The Matrix* to *Jurassic Park*, it has been estimated that sci-fi blockbusters constitute fully one-third of the top fifty biggest film moneymakers in history—and disaster movies don't do too shabbily either. But the relationship between science fiction and actual science is, at best, complicated.

In the sci-fi film genre, there's a kind of unending arms race to achieve ever higher degrees of verisimilitude through ever more stunning computerized graphics and special effects. Reality—or at least the semblance of it—sells. The pursuit of filmic "realism," however, comes about to serve moneymaking goals rather than to satisfy a small, scientifically trained demographic, which may be the only audience group that sees past the veneer enough to contest aspects of it. Consider the dramatic presentation of big-screen dinosaurs in 1993's *Jurassic Park*: Many aspects of their portrayal—the idea that dinosaurs were warm-blooded, for instance—were contested on a scientific level, but viewers thought they were watching the most "realistic" dinosaurs yet depicted. Seeing is believing, especially when the picture is worth millions of dollars.

And so sit the scientists, in the dark, struggling to suspend disbelief but feeling rather ambiguous at best. Yes, science underlies the imaginative leaps of the sci-fi genre, and whenever a big geophysical disaster movie is being made, you can bet that the producers will call on a scientist to consult and lend the film plausibility. Yet scientists remain perennial Hollywood outsiders, always threatening the plot with their pesky emphasis on details. The problem was apparent as early as the 1920s, with Fritz Lang's film *Woman in the Moon (Frau im Mond)*. Lang's science consultant, a rocket scientist named Hermann Oberth, reasonably pointed out that as the moon doesn't have any atmosphere, the characters would have to carry out their roles dressed in spacesuits. Lang replied, "How could one present a love story taking place on the Moon and have the lead characters talk to each other and hold hands through space suits?" Needless to say, there were no spacesuits in the final cut of the film.

To better grasp why science can stumble in its interactions with the entertainment industry, perhaps we need only define our terms. According to Kaplan's Norman Lear Center, the word *entertaining* in its broadest sense simply means "*not boring.*" The entertainment drive, Kaplan adds, is "the imperative to capture and hold attention"; that's how the entertainment industry makes money. To hold attention, successful entertainment requires a strong story line, character development, and high stakes, often of the life-and-death variety. Events unfold at high speed, and there must be serious action or serious drama, not to mention very attractive people.

The problem for science in this context is that the technical facts it furnishes can rarely hold the attention of non-scientists—and anyone who has watched presentations at a scientific conference knows why. That is not to say the goals of science, education, and critical thinking can't also be advanced through entertainment: They can, but only if doing so poses no apparent risk to the product and if the people who matter—writers, directors, producers—see some value in it.

In the many notorious cases where they don't, meanwhile, the most stunning factual and theoretical howlers can occur. It's not merely that you shouldn't look to *Spider-Man* or *X-Men* to learn about the process of mutation, as these films greatly exaggerate it for dramatic purposes. Some of the worst sci-fi and disaster films go much further, seizing upon entirely nonsensical "scientific" premises and then proceeding to come increasingly unhinged as the story advances. In 1997's *Volcano*, for instance, an actual volcano suddenly appears out of the ground to threaten Los Angeles (luckily for L.A., the San Andreas fault, due to its particular tectonic nature, can only produce earthquakes). Then there are the simply excessive made-for-TV disaster movies, like NBC's heavily watched four-hour miniseries *10.5* (about a scientifically impossible earthquake) and CBS's *Category 7* (about a scientifically ridiculous mega-storm). And don't forget the idiotic bloopers, as when a boastful geneticist in 2000's *Red Planet*, showing off his wisdom, names the DNA bases "A, G, T, P." That's right, the actor actually says "*P.*" Didn't the filmmakers see *Gattaca*? A high school science student would have been able to catch that one.

Yet in marshaling scientific complaints against the entertainment industry, it's important to consider what really matters and what doesn't. Any specialist—a historian, say, or an anthropologist—is prone to get ticked off if a film or TV drama makes a mistake about his or her field. And films like *The Core* and *Volcano* probably don't help students or the public understand science—but then, neither are they intended to. So how worried should we really be if an inaccuracy or implausibility slips into a film to serve the plot or to satisfy audience expectations—if, say, *Star Wars* shows fiery explosions in space? Probably not very.

In other realms, though, matters grow more serious. For instance, inaccuracy about medical conditions and treatments can have a real impact on people who get a little too much of their sense of reality from the screen or television. Hollywood's medical plots and subplots are legion—just think of *Grey's Anatomy*, *House*, *Lost* (in which the main character, Jack, is a spinal surgeon), *E.R.*—and these are just prime-time dramas, not movies. These shows pride themselves on their medical realism; the directors and cast of *E.R.*, for instance, make much of the fact that they consult constantly with real doctors. Good for them, but there's much to make up for. For example, one mid-1990s study of television episodes involving CPR found that survival rates were unrealistically high and concluded that "the misrepresentation of CPR on television shows undermines trust in data and fosters trust in miracles."

Perhaps more consequential than scientific inaccuracy, however, is a problem we might call scientist stereotyping. Many groups in society get exercised about how they come across in entertainment depictions, and scientists aren't necessarily the most aggrieved. Still, their complaints have merit. As science-fiction film director James Cameron (*Aliens*, *The Terminator*, *Titanic*, and much else) has observed, the movies generally "show scientists as idiosyncratic nerds or actively the villains."

Cameron's statement finds support in a considerable body of data produced by scholars who have analyzed the treatment of science in film and on television going back several decades. A study of prime-time television programs between 1973 and 1983, for instance, found that one in six scientists was depicted as a villain and that one in ten got killed, both results being the highest rate for any occupational group on television. Such representations have consequences: Researchers who have studied the stereotypical views of scientists held by children even report that when they encounter real-life researchers who visit their classrooms, they think someone's pulling their leg, because the scientists aren't anything like the big-screen version—mean, male, gray-haired, and mad. As one study author explained to the magazine *Nature:* "They might say the person was too 'normal' or too good looking to be a scientist. The most heart-breaking thing is when they say, 'I didn't think he was real because he seemed to care about us.'"

The uncaring scientist, unconcerned about consequences, pursuing knowledge at all costs—this is the ugliest scientist stereotype, and also the most deeply rooted. It hails from a long literary tradition, dating back before Mary Shelley's *Frankenstein* to Goethe's *Faust* and Greek precedents like the story of Prometheus, which depict the search for knowledge as forbidden and dangerous, and leading to disastrous consequences. In this narrative, knowledge leads the scientist to play God, interfere with nature, and attempt to thwart fate by determining who lives and who dies. To know is, in essence, to kill. In film, such depictions go all the way back to Fritz Lang's 1926 classic early sci-fi film *Metropolis*, in which the grotesque mad scientist Rotwang builds an evil robot to resemble the woman he loves, because he can't have her in real life. The paradigmatic modern example of the evil scientist trope is perhaps *E.T.,* in which the scientists, looking like astronauts in their protective gear, want to slice up the cute alien for research.

As we don't see many films about evil literary critics, it's safe to infer there's something about scientists that triggers a particular kind of stereotyping, and that this reflects our society's uneasiness with the power they can sometimes wield. As we saw in the Pluto story that began this book, scientists also make enemies by taking away from people things they cherish—beliefs, settled understandings. Through such actions, they can intimidate and sometimes even enrage non-scientists, and thus they play some role in the construction of their own image.

Without condoning stereotypes, then, or asking scientists to stop being scientists, we might at least suggest some reflection on this fact.

Does Hollywood hate scientists because they're mean and intimidating and spoil all the fun? Probably not. But throughout the industry, there is certainly a sense that science is inimical to storytelling, that it quashes creativity, which must be allowed to breathe. As screenwriter and ScienceDebate2008 founder Matthew Chapman explained about some of his fellow writers: "Among the less talented, there's I think a kind of inherent prejudice against science, because science means being rational, and being rational is considered the opposite of being creative—whereas fantasy, superstition, magic, all of these more child-like ways of looking at life, are somehow thought to be what the creative process is about."

Not everyone in Hollywood shares such sentiments: Many film-makers care about science, as do many writers. But unless they have serious clout within the industry, they may not be able to get their way even if they try. "The person who creates the idea for a story or screenplay is not the person who makes it generally," explains Joe Petricca, a screenwriter and executive vice dean of the American Film Institute. "The person who makes it is about money, business, filling theaters, and sales—and they will exert all the pressure they can to make changes so that it's not so, ah, threatening in that way . . . threateningly intelligent."

The top directors—the Ron Howards, James Camerons, Steven Spielbergs—have enough status in Hollywood to ensure the realization of their artistic visions, which may include the favorable or serious treatment of science (or math, as in Howard's *A Beautiful Mind*). And a number of films have been commended for their plausibility and scientific accuracy, ranging from Carl Sagan's *Contact* (directed by Robert Zemeckis) to the 2003 smash-hit Pixar film *Finding Nemo*, which luxuriated in its sensuous depictions of undersea life and strove, with the help of its science consultant, to achieve the maximum degree of ichthyological realism possible. But when it comes to the sympathetic treatment of science in Hollywood, "the word that comes to mind is *serendipitous*," observes Martin Gunderson, a University of Southern California electrical engineer who has consulted on several films, most notably 1985's *Real Genius*. In other words, it happens—but more because some individual director, writer, or producer cares than because of industry culture.

Hollywood ought to do better, but scientists also need to be realistic about what they can expect. At least at the extreme, we find in the scientists' camp a hectoring annoyance over inaccuracy, the unwillingness or inability to suspend disbelief, and the strange idea that serious science education—instruction about facts and theories—ought to occur through the medium of fictional film and television. These tendencies just won't help in Hollywood interactions. And some scientists will also have to get over the idea that everyone ought to be as captivated by the intricacies of science as they are. "The natural world is fascinating in its own right," Oxford's Richard Dawkins has stated. "It really doesn't need human drama to be fascinating." He even reportedly told the *New York Times* that he wondered why *Jurassic Park* required a cast that included human beings—after all, it already had dinosaurs.

Such science-centrism simply won't work for the broader, non-scientist population. It ignores their compelling need not to be bored. Successes like *March of the Penguins* notwithstanding, most of the time people need to see and hear stories about other people, or about animals that are given human attributes, as in Disney-Pixar films. Dawkins and some other scientists fail to grasp that in Hollywood, the story is paramount—that narrative, drama, and character development will trump mere factual accuracy every time, and by a very long shot. Either science will align itself with these overweening objectives or it will literally get flattened by the drive for profit.

The scientific method, as a process, generally does not make for great storytelling, and particularly not filmmaking. As the late novelist, screenwriter, and movie producer Michael Crichton once put it, there are at least four important rules of movies that don't mesh with the process of research: "(i) Movie characters must be compelled to act. (ii) Movies need villains. (iii) Movie searches are dull. (iv) Movies must move." Crichton argued that real science, with its long, drawnout intellectual processes and frequent dead ends, simply can't be reconciled with such exigencies. "The problems lie with the limitations of film as a visual storytelling medium," he concluded. "You aren't going to beat it."

But maybe scientists can join it? The entertainment industry needs science both to achieve very valuable filmic "realism" and to serve up the most intriguing ideas about the future, in order to furnish compelling sci-fi and disaster story lines. The scientific community should take advantage of this fact and seek out constructive consulting roles within the entertainment industry. It should develop relationships with important players and learn how to serve them to further shared goals, rather than merely issuing criticism and denunciation. Scientists who have done so have frequently reported that the experience was a very enjoyable one, and that they made a real difference. For instance, science consultants working on the 1990s films *Dante's Peak* (about a volcanic eruption) and *Twister* apparently headed off attempts to cast scientific leaders or institutions in the role of out-and-out villains.

Science consultants can have an impact on the scientific content in a film's script, on its set design, on its sound effects. In general, they are invited on board by those at the head of film projects—directors, producers—and their influence is proportionate to the closeness of their relationship to that leader. By the time a science consultant arrives on the scene to work on a project, many things such as plotline, cast, and budget are usually already agreed upon, and a script has likely been written, at least in draft form. Given all of this, any effective science consultant or adviser will be acutely aware of the realities and constraints of filmmaking and will work with them, rather than trying to overturn them.

It's worth taking a chance on collaborations of this kind because the stakes are extremely high. We're talking about perhaps the single most influential slice of the media, an arena in which a success can have a dramatic impact.

To see as much, one need only consider Al Gore's *An Inconvenient Truth*. It was a very unconventional film—a documentary centered on a PowerPoint presentation—but something in the zeitgeist just clicked, perhaps because the messenger was the man who should have been president. Thus, the film's smash success not only transformed Gore's image and influence but also helped to move the global warming debate onto a new footing. Gore's subsequent garnering of an Oscar (again bestowed by Hollywood) and a Nobel Peace Prize (not bestowed by Hollywood, but certainly very much influenced by it) only broadened the impact further. The stars aligned, and a Hollywood film had a massive impact in translating science's relevance to policymakers and the public.

Hollywood can also shape and direct the pursuit of scientific careers. The success of CBS's television series *CSI: Crime Scene Investigation*, which glamorizes and dramatizes forensic science, led to a much-noted growth of student interest in the field and the proliferation of university programs to service that demand. This happened despite the fact that real forensic work is hardly as exciting and dramatic as it appears on TV, and few forensic scientists are as hot.

Perhaps these examples show why we badly need more go-betweens to link the science world and the entertainment industry. Luckily, the science community has begun to recognize this and work in a concerted way to promote it. The U.S. National Academy of Sciences—whose members once spurned Sagan—recently launched a project entitled the Science and Entertainment Exchange, which aims to "facilitate a valuable connection between the two communities." This included the opening of a permanent National Academies office in Los Angeles to "make introductions, schedule briefings, and arrange for consultations to anyone developing science-based entertainment content." This is a new initiative, so one cannot yet judge its success, but it is precisely the type of outreach the scientific community should be engaging in if it wants to take advantage of the immense opportunities afforded by the mass medium of film.

Sagan in fact epitomizes both the promise and the perils of scientists working with Hollywood. His 1997 film *Contact*, based on his novel of the same name, strove to preserve scientific accuracy and plot plausibility. It also depicted the main character, astrophysicist Ellie Arroway (played by Jodie Foster), as a positive role model, a visionary who overcomes numerous obstacles to pursue her quest of determining whether extraterrestrial life really does exist and is trying to reach us.

Contact didn't do badly. It earned $171 million worldwide, nearly doubling the $90-million production budget. But it was also competing with two considerably more brainless alien-related sci-fi movies, 1997's *Men in Black* and 1996's *Independence Day*, both of which made a great deal more money ($589 million and $817 million, respectively). And that unfavorable financial comparison, more than anything else, dramatizes the incredible challenge science faces in Hollywood. If films that strive to balance scientific plausibility with a compelling story line don't succeed at the box office, the road will be tougher for future efforts. Luckily, scientific stinkers don't necessarily make the biggest fortunes, either. When you compare two dueling 1997 volcano movies, the more plausible *Dante's Peak* and the much less scientifically serious *Volcano*, the former did more business ($178 million worldwide) than the latter ($122 million worldwide).

Such figures provide hope that Hollywood could play a role that serves the interests of the entertainment industry, science, and society alike by helping us grapple with the future we are hurtling toward. The ideal synergy would occur if more sci-fi and disaster plots took as their basis the problems that we really need to worry about, and dramatized them compellingly. There have been some excellent, and successful, examples of such films in the relatively recent past, and in closing, let's survey them.

Among science-related topics, human cloning and genetic engineering—issues that are already raising ethical concerns for us now and that will likely raise a great many more in the future—seem to generate some of the very worst filmmaking (as anyone who has seen *The Sixth Day* or *Godsend* or *The Island* can affirm). But 1997's *Gattaca*, starring Ethan Hawke and Uma Thurman, represents an exception, a counterexample showing how filmmaking can help audiences contemplate the challenges that new discoveries could pose down the road.

In the futuristic film, Hawke plays Vincent Freeman, a so-called In-Valid because his parents did not opt, before his birth, to have a child whose genetic defects had been weeded out. This puts Vincent into a lower genetic caste, limiting his ability to achieve his dream of flying into space to Titan, the largest of the Saturnian moons. The film details how Vincent overcomes the obstacles posed by his genetic "limitations"—because in the end, genes aren't everything. In laying out such a story line, *Gattaca* undercuts a kind of unthinking proposition that we've been fed all too often, which one critic called "genetic determinism." In many Hollywood films, the techniques of biotechnology and genetic manipulation appear all-powerful and perfect; in the Arnold Schwarzenegger film *The 6th Day*, for instance, it's no big deal to clone an exact replica of a human being. *Gattaca* challenges such questionable presumptions and constructs a far more realistic scenario in which the techniques of genetic manipulation have become unevenly distributed in society, creating considerable inequities, while at the same time, some genetically "limited" people overcome the hurdles they start out with even as some genetically "gifted" ones fail to achieve their potential.

Another film that tried to prepare us for the future—the 2004 blockbuster *The Day After Tomorrow*, which grossed a very impressive $542 million worldwide—focused on global warming. Unfortunately, on a scientific level the film's plot is risible: It depicts climate change as the trigger for an instantaneous new ice age (huh?) and shows the world facing a massive, coordinated assault by upside-down, freezing hurricanes that somehow travel over land. Still, the movie also features a scientist (played by Dennis Quaid) in the lead protagonist role and contains a considerable amount of accurate dialogue and even some speeches about climate science—before the improbable lunacies begin, anyway. Virtually all of the film's top characters are scientists and are treated sympathetically; and the message, inaccuracies aside, is that global warming is a problem that we can't delay addressing, lest very bad things happen—so let's listen to the scientists before it's too late.

It would surely have been possible to make a global-warming-related disaster movie with a much closer connection to the actual risks posed. Still, it appears *The Day After Tomorrow* did have an important effect on its audiences. According to one study, those who had seen it were significantly more worried about global warming than those who had not and were significantly more convinced that global warming could trigger specific weather and climatic impacts (including, unfortunately, the idea of a new ice age caused by an ocean-current shutdown).

Contact, Gattaca, and *The Day After Tomorrow* all demonstrate the potential for collaboration between the world of science and the world of Hollywood, and suggest such interactions could be mutually beneficial, at least if those who care about science take up the challenge of connecting in a more positive way with the film and television industry.

That challenge must be met quickly, because the enemies of science certainly recognize the medium's potential as a tool of propaganda. For example, the 2008 right-wing documentary *Expelled!* features the comedian Ben Stein in a dishonest look at the evolution controversy that slanders the scientific community for intolerance toward religion and the quashing of anti-evolutionist dissent. Throwing the kitchen sink at evolution, Stein not only charges that pro–intelligent design scientists have been repressed on university campuses by dogmatic Darwinists, but even preposterously blames Charles Darwin and his work for the Holocaust.

This is all quite inaccurate, even ludicrous; but the message conveyed by *Expelled!* about the evils of the scientific community has reached a lot of people, because it debuted at 1,000 theaters across the country. It ultimately earned over $7 million, making it the fifth-highest-grossing political documentary ever.

Expelled! represents a cultural warning sign that should not go unheeded. Film and television are massively powerful media and can be used as damaging weapons. Scientists must learn how to wield them as well, and for more virtuous purposes. They've been in the dark for far too long.

CHAPTER 8

Bruising Their Religion

IN SUMMER 2008, A UNIVERSITY OF CENTRAL FLORIDA STUDENT NAMED Webster Cook removed a communion wafer from Catholic Mass and took it home with him—according to one news report, holding it "hostage." Cook seems to have been unhappy about the fact that his state-sponsored university funded religious groups; he also claimed he simply wanted to show the wafer to a friend. But as the Eucharist is one of the seven sacraments in the teaching of the Catholic Church and considered the literal presence of Christ, his action outraged many Catholics, sparking a flurry of media coverage and even, apparently, some death threats. As the controversy escalated, Cook soon returned the wafer in a Ziploc bag. There things might have ended, had the popular atheist science blogger and University of Minnesota professor Paul Zachary ("PZ") Myers not gotten involved.

Myers was staggered and disgusted by all the hoopla over a "frackin cracker." So on his widely read weblog Pharyngula, he asked if anyone out there could "score *me* some consecrated communion wafers?" and promised, in return, "profound disrespect and heinous cracker abuse, all photographed and presented here on the web." Whether he was initially joking or not, before long the matter became quite serious, and Myers—himself now under fire from Catholics—carried out the deed:

I wasn't going to make any major investment of time, money, or effort in treating these dabs of unpleasantness as they deserve, because all they deserve is casual disposal. However, inspired by an old woodcut of Jews stabbing the host, I thought of a simple, quick thing to do: I pierced it [the Eucharist] with a rusty nail (I hope Jesus's tetanus shots are up to date). And then I simply threw it in the trash, followed by the classic, decorative items of trash cans everywhere, old coffee-grounds and a banana peel. My apologies to those who hoped for more, but the worst I can do is show my unconcerned contempt.

Accompanying these words, a photograph of the "Great Desecration" appeared on Myers's blog, showing that along with the trashed Eucharist, he had also disposed of pages from the Qur'an and Richard Dawkins's best-selling 2006 book, *The God Delusion*—just to prove nothing is sacred.

Myers's public desecration generated a global outcry. The Catholic League histrionically demanded that the University of Minnesota take disciplinary action against him, and threatening e-mails arrived in droves. Legally, of course, Myers was exercising his freedom of expression. He should not have been fired or disciplined, and thankfully wasn't.

Nonetheless, Myers's actions were incredibly destructive and unnecessary. He's a very public figure: His blog often draws over 2 million page views per month. It was dubbed the top science blog by *Nature* magazine in 2006 and appears on the ScienceBlogs.com network; its author also writes for the popular-science magazine *Seed*, whose motto is "Science Is Culture." Yet Myers's assault on religious symbols considered sacred by a great many Americans and people around the world does nothing to promote scientific literacy; rather, it sets the cause backward by exacerbating tensions between the scientific community and many American Christians. To religious fundamentalists, who already nourish plenty of suspicion toward mainstream science—and who promote the false and pernicious idea that the faithful can't accept such core scientific findings as evolution—Myers is a walking, blogging justification for their confusion and misdirected antipathy. Religious moderates will scarcely be more receptive: The Catholic Church (whose sacrament Myers defiled) is not even opposed to evolution (which Myers teaches). If episodes like the "Great Desecration" succeed at anything, it will be the fostering of greater alienation and divisiveness between two groups that have had more than enough battles over the centuries—scientists and religious believers.

Myers is certainly not alone. In recent years a large number of "New Atheist" voices have arisen, a movement that originated in significant part in response to the religiously driven massacre committed by al Qaeda on September 11, 2001. The writers Sam Harris, Richard Dawkins, Christopher Hitchens, and Daniel Dennett are generally considered the "big four" (or if you prefer, the "four horsemen") of new atheism. They're hardly a monolithic group—Harris in fact rejects the atheist label. But the broad tenor of the movement they've impelled is clear: It is confrontational. It believes religious faith should not be benignly tolerated but, rather, should be countered, exposed, and intellectually devastated.

The most outspoken New Atheists publicly eviscerate believers, call them delusional and irrational ("demented fuckwits," as Myers put it in the Webster Cook case), and in some cases do not spare more liberal religionists, or even more conciliatory fellow scientists and atheists, from withering denunciation. Sam Harris questions the very notion of tolerating religious moderates; he thinks they merely enable extremists. For Richard Dawkins, meanwhile, those who do not criticize religion but still want to defend the teaching of good science in schools fall into the "Neville Chamberlain school of evolutionists" and the "appeasement lobby."

If the goal is to create an America more friendly toward science and reason, the combativeness of the New Atheists is strongly counterproductive. If anything, they work in ironic combination with their dire enemies, the anti-science conservative Christians who populate the creation science and intelligent design movements, to ensure we'll continue to be polarized over subjects like the teaching of evolution when we don't have to be. America is a very religious nation, and if forced to choose between faith and science, vast numbers of Americans will select the former. The New Atheists err in insisting that such a choice needs to be made. Atheism is not the logically inevitable outcome of scientific reasoning, any more than intelligent design is a necessary corollary of religious faith. A great many scientists believe in God with no sense of internal contradiction, just as many religious believers accept evolution as the correct theory to explain the development, diversity, and inter-relatedness of life on Earth. The New Atheists, like the fundamentalists they so despise, are setting up a false dichotomy that can only damage the cause of scientific literacy for generations to come. It threatens to leave science itself caught in the middle between extremes, unable to find cover in a destructive, seemingly unending culture war.

Of course, the New Atheists aren't the *origin* of the cleft between reli-gious and scientific culture in America—they're more like a reaction to it. They're responding to a tension that has pervaded American history and that broke out openly and perhaps most dramatically during the 1925 Scopes "monkey trial."

Ever since evangelicals reemerged as a political force in the 1970s, faith and science have been pitted against each other repeatedly in con-troversies over topics ranging from abortion to stem cell research to the teaching of evolution in American schools. The emergence of the highly organized intelligent design movement in the 1990s, as well as the excruciatingly anti-science presidency of George W. Bush, precipi-tated a new wave of attention to the fissure between science and faith in America. We've already seen that an alarming percentage of our citi-zens (46 percent) subscribe to young-Earth creationism, the scientifi-cally untenable idea that God created the earth and everything on it within the last 10,000 years. And as with evolution, so with the Big Bang: Less than 40 percent of Americans can correctly answer the ques-tion "The universe began with a huge explosion [true or false]." In Japan and South Korea, by contrast, the number is over 60 percent.

Meanwhile, there's no question that America's scientific community is far more secular in outlook than the rest of the nation. A 2007 study revealed that whereas 52 percent of scientists at twenty-one leading U.S. academic institutions claimed to have no religious affiliation, that was true of just 14 percent of the broader U.S. public. And whereas 14 percent of Americans self-identify as "evangelical" or "fundamentalist," fewer than 2 percent of the surveyed scientists did. The study also revealed that far more than the general population, scientists tend to come from liberal or nonreligious family backgrounds. In fact, those scientists in the survey who professed religious beliefs tended to have grown up with them; childhood upbringing was a central factor in separating religious and non-religious scientists. The authors of the study concluded, "While the general American public may indeed have a less than desirable understanding of science, our findings reveal that academic scientists may have much less experience with religion than many outside the academy."

This large divergence of assumptions and backgrounds manifests itself in what is by far the most religiously contested scientific issue in the United States, the teaching of evolution. On this subject, many scientists have a strong and almost instinctual inclination to exhaustively refute the claims made by creationists of various stripes: They feel compelled to show why they're wrong, how they misrepresent science, that they violate scientific canons and norms, and so on. Accordingly, the specific claims of so-called scientific creationism—that the earth is no more than 10,000 years old; that its geological features, such as the fossil record and the Grand Canyon, were carved by Noah's flood; that humans coexisted with dinosaurs; and so on—have been wholly dismantled by scientific experts. The same goes for various claims made by the newer intelligent design movement, such as the assertion that certain cellular components are "irreducibly complex," and therefore suggest the hand of an intelligent designer (God).

The scientific case for rejecting such bad science (or non-science) is indisputable. But that doesn't make it persuasive to creationists or other religiously motivated evolution skeptics. Although anti-evolutionist leaders may dress up many of their claims in scientific trappings, the vast majority of their followers aren't really operating on that level, to the continual befuddlement of some scientists, who keep laying out the facts but seeing no one swayed who wasn't already on the pro-evolution side. "The appeal of creationism is emotional, not scientific," writes Kenneth Miller, a Roman Catholic biologist, leading evolution defender, and author of *Finding Darwin's God*. Creationists are driven to attack evolution because they fear it will undermine their religious culture, which for many is the essential organizing principle of their lives and the lives of their children. Abrasive atheism can only exacerbate this anxiety and reinforce the misimpression that scientific inquiry leads inevitably to the erosion of religion and values.

This observation doesn't make those religious conservatives who reject science, who constantly battle over the teaching of evolution, a bunch of innocents—far from it. They do their devastating part to keep our culture from embracing science as it should. But it does put the community of science on notice that whatever external problems we face, we're also troubling our own house. To further the cause of scientific literacy, we need a different, and far more sympathetic, approach, one that's deeply sensitive to the millions of religious believers among our citizenry.

Fortunately, this prescription isn't merely political or strategic in nature; it's substantive and entirely based on reason. For it turns out that the New Atheists are quite incorrect about the relationship between science and religion on multiple levels, from the historical to the philosophical. Not only are they causing a great deal of divisiveness, but they're doing so on the basis of what are, at best, questionable premises.

At a recent conference at City College in New York City, an audience member asked a panel of Nobel laureates whether a true scientist could also believe in God. Chemist Herbert Hauptman answered with a definitive "No!"—reasoning that quality science and supernatural beliefs are irreconcilable and adding that such beliefs are "damaging to the well-being of the human race." Yet historical scholarship on the complex interactions between science and religion contradicts Hauptman's simplistic assertion. A great many leading lights of the scientific revolution and the Enlightenment—Nicolaus Copernicus, Francis Bacon,

René Descartes, Johannes Kepler, Galileo Galilei, Isaac Newton, Robert Boyle—were distinctly religious and viewed science as a better means of understanding God's creation and the laws governing it.

Granted, there have also been many episodes of conflict between science and faith: The stories of the sixteenth-century Copernican Giordano Bruno (burned at the stake by the Roman Inquisition for his unorthodox views, including the contention that the universe is infinite and there are many worlds) and Galileo (censored, forced to recant, and committed to house arrest by Pope Urban VIII for defending Copernican heliocentrism) certainly come to mind. Two influential late-nineteenth-century books—1874's *History of the Conflict Between Religion and Science,* by New York University professor John Draper, and 1896's *A History of the Warfare of Science with Theology in Christendom,* by Cornell University president Andrew Dickson White—marshaled such examples into a formal science-religion "conflict" narrative, which the New Atheists carry forward today.

And thus while science and faith are not mutually exclusive, it's certainly fair to say the two have posed tremendous challenges for each other over the course of history. Beginning with the scientific revolution in the sixteenth and seventeenth centuries, extending through the cultural thunderclap that was the publication of *On the Origin of Species* in 1859 and up to the present day, science has continually usurped terrain previously occupied by Christianity. It has established its own intellectual independence and cultural authority; it has triumphed repeatedly with regard to numerous factual claims about the world and the workings of nature. Many scripturally based "scientific" assertions—the statement in Psalm 104:5 that "Thou didst set the earth on its foundations, so that it should never be shaken" or in Ecclesiastes 1:5 that "the sun rises and the sun goes down, and hastens to the place where it rises," to name a few—have been undermined by science and become impossible for critically thinking people to take literally.

Increasingly, science has offered explanations for aspects of existence or reality that were once described as miracles or the product of divine intervention. Over time, diseases have come to be understood as naturally caused, not to be cured by exorcism or prayer but by medical treatments. Similarly, an awe-inspiring phenomenon like lightning—once assumed to be divinely impelled—was recognized to be a natural occurrence. This new understanding, in turn, suggested how to protect human structures from its ravages—and thus Benjamin Franklin invented the lightning rod.

The Darwinian revolution further advanced this trend of desacralizing nature. Throughout the first half of the nineteenth century and especially in England, many thinkers still looked to the natural world for "scientific" evidence of divine handiwork. They held that its intricate wonders, particularly the seemingly miraculous contrivances that allow living organisms to thrive, such as the eye, constituted proof of a designer's active hand. Yet Darwin devastatingly unseated the idea that living things provide evidence of the divine in their bodily structures or attributes; instead, a directionless process acting over long periods of time—natural selection—could account for the diversity of life as well as the adaptive efficiency of its organs and organisms. Species hadn't been created at fixed moments in time by an intelligent agent, and their identities weren't static. They did not constitute the obvious evidence for God's existence that so many had assumed; once again, religion would have to retreat in the face of scientific advance.

The Darwinian revolution, and the considerable religious resistance it triggered, certainly upped the intensity level for conflicts between science and religion. But again, that does not mean acceptance of evolution and belief in God are incompatible. Even at the time, many religious thinkers, including major Anglican clergymen such as Charles Kingsley and Frederick Temple, had no problem adapting to Darwin's ideas.

Darwin personally struggled with the science-religion question in his own life, torn between his devoutly Christian wife, Emma, and anti-clerical combatants like his brawling disciple, Thomas Henry Huxley. But he was clear about the nature of science: "The more we know of the fixed laws of nature," he wrote in his *Autobiography*, "the more incredible do miracles become." Scientists today might not yet understand every detail of nature's causal chain, but that's why they conduct further research rather than expecting religion to fill any of the remaining holes. We've come a long way since the days when theological and scientific reasoning regularly commingled.

Yet religion has come a long way, too: The history of world religions is one of continuous adaptation, not only to science but to many other developments as well. Large numbers of religious people today reject the idea that the Bible must be read literally and every word accepted as eternally true, or that the Ten Commandments represent a complete prescription of ethical behavior. Hordes of believers worldwide are able to reconcile ancient faiths with modern developments that range from the emergence of feminism, to changing understandings of human rights, to racial equality, and yes, to scientific progress as well. This is not to deny that science has posed considerable challenges for the Judeo-Christian tradition; it is simply to say that there are a wide variety of potential responses to these challenges. The official position of the National Academy of Sciences and the American Association for the Advancement of Science is that faith and science are perfectly compatible. It is not only the most tolerant but also the most intellectually responsible position for scientists to take in light of the complexities of history and world religion.

The problem with the New Atheism, however, isn't just that it's divisive or historically incorrect about the relationship between science and religion. It's also misguided about the nature of science. To understand why we say this, it's essential to understand the scope and evidentiary basis of scientific inquiry.

Modern science relies on the systematic collection of data through observation and experimentation, the development of theories to organize and explain this evidence, and the use of professional institutions and norms such as peer review to subject claims to scrutiny and ultimately (it is hoped) develop reliable knowledge. A core principle underlying this approach is something called "methodological naturalism," which stipulates that scientific hypotheses are tested and explained solely by reference to natural causes and events. Crucially, methodological naturalism is *not* the same thing as philosophical naturalism—the idea that *all* of existence consists of natural causes and laws, period. Methodological naturalism in no way rules out the possibility of entities or causes outside of nature; it simply stipulates that they will not be considered within the framework of scientific inquiry. As Michigan State University science philosopher Robert Pennock put it in his testimony at the Dover, Pennsylvania, evolution trial in 2005, we're talking here about a "method that constrains what counts as a scientific explanation." It is very clear today, Pennock noted, that "when one does science, one is setting aside questions about whether the gods or some supernatural beings had some hand in this."

The critical point is that such naturalism, being merely methodological in nature, is not a claim about the fundamental reality of the world. And it certainly is not atheism. Rather, it's simply a rule that is justified on pragmatic grounds by the dramatic success it has facilitated in the application of scientific knowledge to real-world situations. Anti-evolutionists have long sought to slander the scientific enterprise by claiming it's inherently atheistic, but in so doing, they misunderstand both the history and the philosophy of science. Or as Pennock amusingly puts it: "Science is godless in the same way that plumbing is godless."

But much like anti-evolutionists do, the New Atheists often seek to collapse the distinction between methodological and philosophical naturalism. In *The God Delusion,* for instance, Richard Dawkins makes the dubious claim that the existence of God is, as he puts it, "unequivocally a scientific question." Quite a lot of philosophers—and scientists—would disagree. It is one thing to say that scientific norms and practices preclude ascribing any explanatory force to God in, say, the movement of atoms, or the functioning of DNA. It's quite another to say they entirely preclude God's existence. In rejecting God or any other supernatural entity, Dawkins is taking a *philosophical* position. He has every right to it, of course. But to pretend that there is something inherent in science that requires him to do so is an intellectual error at best—and at worst, a nasty bullying tactic.

The reassuring fact is that despite the shrill battles between the antiscience fundamentalists and the New Atheists in recent years, most

Americans seem to understand that science and religion are perfectly compatible. A 2006 study sponsored by the Faith and Progressive Policy Initiative of the Center for American Progress, for instance, found that 80 percent of respondents agree that "faith and science can and should coexist." A great many religious organizations in America also uphold the principle of compatibility. All of the following, for example, accept the teaching of evolution: the Roman Catholic Church, the General Convention of the Episcopal Church, the Central Conference of American Rabbis, the United Church of Christ, the American Jewish Congress, the United Presbyterian Church USA, and the United Methodist Church. Or consider the Clergy Letter Project, which as of this writing had drawn the signatures of nearly 12,000 Christian religious leaders who "believe that the timeless truths of the Bible and the discoveries of modern science may comfortably coexist" and that "the theory of evolution is a foundational scientific truth, one that has stood up to rigorous scrutiny and upon which much of human knowledge and achievement rests."

The American scientific community gains nothing from the condescending rhetoric of the New Atheists—and neither does the stature of science in our culture. We should instead adopt a stance of respect toward those who hold their faith dear, and a sense of humility based on the knowledge that although science can explain a great deal about the way our world functions, the question of God's existence lies outside its expertise. Our previous—and now unfortunately deceased—top public science communicators, Carl Sagan and Stephen Jay Gould, understood this. They brought people together to appreciate science and recognized that with their stature came a moral and intellectual responsibility to help our society deal with difficult issues—among them the apparent conflicts between science and religion—through moderation and accommodation.

Long before Richard Dawkins was denigrating "the weakness of the religious mind" (as he does in *The God Delusion*), Carl Sagan subjected all the standard arguments for God's existence to strong scrutiny in his 1985 lecture series, "The Varieties of Scientific Experience: A Personal View of the Search for God." Sagan himself was not convinced by any of them; he remained agnostic. But he also treated the subject of religion respectfully and acknowledged its many benign and helpful functions. And he called for humility, and for mutual respect, as a precondition to dialogue.

Now more than ever, we need to heed this lesson, end the current polarization, and restore a broader sense of compatibility between science and religion. There's no time to waste: In the coming years, we're likely to see many scientific discoveries that raise troubling ethical questions and spark new conflicts. Neuroscience, for instance, is providing an increasingly naturalistic picture of human consciousness, as brain imaging has led some to suggest we are nothing more than electrical impulses and proteins. This reductionist perspective will, assuredly, challenge fundamentalist religion and widely held notions about humanity—free will, the concept of the soul. In fact, it's already happening: As we completed this chapter, *New Scientist* magazine ran an article, entitled "Creationists Declare War over the Brain," detailing how the arguments of intelligent design are now being imported into the neuroscience arena to challenge the idea that we're just matter in motion.

As such battles erupt, it will be crucial for people of faith to feel they can engage in an open dialogue with people of science to help understand these emerging technologies and ideas, and to find ways of integrating them into their worldview. And who knows? Scientists might get something out of the exchange as well. After all, the faithful strove to understand the world in their particular way for several millennia before modern science got into the game. Although we no longer turn to them for explanations of workings of nature—and shouldn't—they have a vast store of knowledge about what it takes to motivate people, create community, and bring about social change. These are lessons that scientists, and people of reason, could do worse than to heed.

PART 3

THE FUTURE
IN OUR BONES

Funding Executive: Your proposal seems less like
science and more like science fiction.
Ellie Arroway: You're right, it's crazy . . . all I'm asking is for you to have just the tiniest bit of vision.

—CONTACT, 1997

CHAPTER 9

The Bloggers Cannot Save Us

WITH JUST DAYS TO GO UNTIL VOTING CLOSED, THE 2008 WEBLOG AWARDS— an annual online popularity contest in which nearly 1 million voters pick their favorite opiners across forty-eight topic categories—featured a tight race for Best Science Blog. The two leading contestants, nearly neck and neck: PZ Myers's Pharyngula, the online clearinghouse for confrontational atheism, and a blog called Watts Up With That, written by former TV meteorologist Anthony Watts, a skeptic of human-caused global warming.

With the vote running close and Watts leading, in early January 2009 the standard e-mails went out in science circles—"Vote for PZ, we can't let the climate denier win"—yet it was all for naught. In the end, Watts Up With That defeated Pharyngula by a vote of 14,150 votes to 12,238. And for some Myers voters, it may have been a devil's choice anyway. Sure, they could have voted for a blog that discusses science accurately and without anti-religion polemicism—like, say, the global warming site Real Climate or Phil Plait's Bad Astronomy, which were also in the running—but it was clear from early voting trends that such outlets had virtually zero chance of winning. They lagged much too far behind the leaders. The real popularity contest came down to the religion basher and the misinformation machine, and that speaks volumes about the form "science" commentary now takes on the Internet.

As traditional science journalism gasps for each breath, there's no doubt that vibrant science conversations have migrated to the Web and are occurring through a new form of citizen-journalist media—blogging. Although the term "blog" may once have evoked images of Steve Urkel and other stereotypical geeks, weblogs have entered mainstream consciousness in recent years and have exploded in popularity and readership. As of 2008, 93 percent of the leading 100 newspapers in the United States had blogs written by their reporters, and the largest blogs, such as the Huffington Post and Daily Kos, now rival many traditional media outlets in their online readership.

The scientific community came relatively late to the blogosphere, but in recent years online science commentary and opinion have mushroomed. It's estimated that there are some 1,000 science blogs in existence, and that's probably very conservative. Science blogs often focus on hot-button topics such as vaccination, the teaching of evolution, and the politics of climate change, and also frequently devote considerable time to parsing new research findings. Whatever their chosen fields, science bloggers often approach them with considerable verve and even attitude, so it's no wonder they have often been touted as (and act like) a revolutionary solution to the long-standing science-communication problem.

Indeed, science blogs aren't just personal endeavors any longer; they're a business. Some prominent hubs such as RealClimate.org remain independent, but conglomerate sites like ScienceBlogs.com and Discover Blogs (where we hang our hats) have grown increasingly dominant. Owned by Seed Media Group, which also publishes *Seed* magazine, ScienceBlogs now aggregates over seventy individual blogs, carries online advertisements from major corporations like Honeywell, and even has a life-science blog sponsored by the biotech firm Invitrogen.

And the growth of blogging is only one example of the changing science-media landscape. In all sorts of other ways as well, the Internet has become the go-to place for science: According to the National Science Foundation, it now ranks second only to television among the leading sources of information about science for the average citizen and is increasingly leaving other, older sources in the dust. In particular, when Americans want to find information about a specific scientific topic, they now go to the Web far more often than they go to the library—or even to their own bookshelves—to pluck out a reference book. The unavoidable question is whether this is a good thing. Although it *seems* like the Internet should be a very hospitable environment for scientists—after all, they invented the thing—in reality it's less than clear whether the great democratizing force of the World Wide Web will revitalize American scientific culture, or simply degrade it still further.

The problem with the Internet is obvious to anyone who has ever used it: There's tons of information available, but much of it is crap. This is as true for science as for any other area; the Web empowers, but it empowers good and bad alike. Much like cable television does for its viewers, the Internet offers users a huge array of places to go. Yet this very feature entails the option to avoid anything serious or informative. It also lets people join up with those who think just like they do, in back-scratching communities that rarely encounter anything challenging or unexpected.

When it comes to science blogging in particular, it has been suggested that the medium allows practitioners to bridge the traditional "two cultures" divide, yet the foregoing line of thinking implies a darker outcome. If bloggers who write about science are mostly writing for others who already care about it, we might expect to see the reinforcement of some tendencies that have not necessarily helped science when it intersects with broader society: a sense of intellectual superiority; a belittling of politicians, the media, and the public; and attacks on religion. Meanwhile, we would also expect to see anti-science forces, such as global warming and evolution deniers, establish their own Internet hubs, talk to their own friends and pat *them* on the back, and follow the same pattern. And the 2008 Weblog Awards seem a perfect confirmation: Anti-religion and anti-science polemicism are *both* very popular, and *both* have devoted audiences.

That's not to say science blogging is all rotten—far from it. We feel the vibrant online conversation about science has many benefits, and we continue to participate in it daily. And we occasionally swear in wonder when the blogosphere flexes its muscles and brings the mainstream media—what bloggers call the "MSM"—to its knees.

While this book was going to press in March 2009, for instance, a mob of science-oriented, political, and environmental bloggers eviscerated the right-wing newspaper columnist and pundit George Will over a thoroughly inaccurate column about global warming. And when the *Washington Post* failed to publish a correction to Will's howling factual errors—for example, he made the soundly refuted claim that during the 1970s, the scientific community had reached consensus about the dangers of "global cooling"—they turned their fire on the paper as well. It was in many ways a powerful spectacle, a demonstration of just how incisive bloggers can be—especially when it comes to getting the facts straight about some of the most important (and complicated) scientific issues that face us—and how threatening they are to the ailing mainstream media.

As this example shows, it's indisputable at this point that blogs *matter*, in large part because of their ever growing influence on the traditional press. It's no accident that big journalism outlets like the *New York Times* are either acquiring or simply launching outright the most significant science-related blogs, annexing some of the most influential voices in the new media and merging them with the old.

And science blogs don't merely help the science crowd influence and lobby the older media; they also facilitate high-level discourse among experts, allowing the advancement of knowledge and understanding, and even the framing of new research questions, to proceed considerably faster than they did before the advent of the Internet. A perfect example is the dialogue that occurs on RealClimate, where leading climate scientists regularly read and participate almost as if they're at an online version of a scientific conference, and comment threads include hundreds of high-level, astute contributions.

Blogging helps provide a more democratized, user-friendly, and open-access dialogue about science, especially at a time when the nonsubscriber price to read a single article in *Nature* is $32. And it affords the opportunity to bring audiences along to remote locations where scientific research is taking place. At a time when mainstream media coverage of science is declining and budget slashing cuts back on reporter travel, scientists still head into the field regularly, and now they can report back online. Many do.

Finally, and perhaps most important: Along with online social-networking sites like Facebook and grassroots sites like MoveOn, blogs are an incredibly powerful tool for political organizing—for quickly mobilizing like-minded individuals behind a cause and constantly feeding them informational updates that keep that cause at the front of their minds. A classic example occurred with ScienceDebate2008: Although the mainstream media largely ignored the initiative, the formation of a "blogger coalition" and Facebook group allowed it to rocket rapidly around the Internet, which led to a large volume of endorsers of the cause in the science world. This is the kind of example that must be built upon, where blogs can probably do the most for science in its intersection with the public.

Let no one say we don't give science blogging its due. Still, how much time should the scientific community be spending online? The audience for science-specific blogging is probably both narrower and far more specialized than the audiences for television, film, or major newspapers. And in the vast space of the Internet, it's unlikely that anyone who is not already a science aficionado is going to spend much time with science blogs, except perhaps by stumbling over them by accident or while researching a topic.

Blogging also consumes a vast amount of time due to the speed of dialogue and the demands of "feeding the beast," a set of pressures that inevitably leads to much quick writing and posting rather than deep, sustained thought, and that favors polemicism over nuance. This means science blogging can rarely serve as a real substitute for in-depth, considered, professional science journalism of the sort that is now in demonstrable decline—the kind of time-consuming writing that canvasses researchers, peruses the literature, and truly penetrates into where science is headed and why it matters.

The single biggest blogging negative, however, is the grouping together of people who already agree about everything, and who then proceed to square and cube their agreements, becoming increasingly self-assured and intolerant of other viewpoints. Thus, blogging about science has brought out, in some cases, the loud, angry, nasty, and profanity-strewing minority of the science world that denounces the rest of America for its ignorance and superstition. This ideological content, which inflames audiences, is often the most likely to draw attention outside of the science-centric blogosphere—meaning that out of the many contributions made by science blogging, the posts that non-scientists (or people who don't follow science regularly) will most probably come across are those skewering religion.

It is no accident, then, that PZ Myers's Pharyngula is such a popular science blog, though its content is hardly limited to science. Myers currently receives more than 1.5 million unique visitors each month and approaches 3 million page views, with nearly 2 million unique visits during the month that he posted an image of his desecration of a communion wafer. This may have been the single biggest opportunity yet for science content to break out of the science corner of the blogosphere, but it was likely also the most alienating one.

Myers's opponent in the 2008 Weblog Awards contest, of course, was arguably even worse, and certainly less helpful to the cause of scientific literacy in America. Anthony Watts is an extremely popular blogger, drawing hundreds of comments per post and well over half a million visitors per month. Yet his blog contains highly questionable information—presented very "scientifically" of course, replete with charts and graphs—but all directed toward the end of making the scientific consensus on human-caused global warming seem faulty (in fact, it's extremely robust). A particular delight of the blog: hyping individual winter-weather events as if they have something to do with refuting global warming trends, a basic error of statistical reasoning.

It's not just in the global warming realm of the Internet that misinformation thrives; the modern day vaccine skeptic movement, a particularly dangerous outgrowth of anti-science sentiment, also owes much of its present force to the Web. And though there are excellent blogs that debunk anti-vaccine claims—such as Respectful Insolence—many other high-traffic sites prop them up again, including the Huffington Post.

The point is that it's a Wild Wild West out there, yet another incarnation of the free market applied to the communication of science. And though all the chatter and back-and-forth may produce strong feelings and intensely committed communities fired up about particular issues—on either side of them—it's not unifying us behind science or bringing it back to the center of our culture. Science blogs, at least as they currently exist, don't seem to present the kind of intervention necessary to restore science's stature and relevance in our national dialogue.

But will blogs always be as they "currently exist"? Almost certainly not. If we've learned anything from the burgeoning online community, it's that this new mode of communication is constantly developing and changing. The blogosphere has barely entered its adolescence and will continue to mature through technological innovation and cultural change—as will the Internet.

What comes next? At a time when streaming video over the Web is becoming ever higher in quality, technophiles are salivating over the day when the Internet and television ultimately merge into a single medium—just as soon as someone invents the right platform to integrate the two (and subsequently grows very wealthy). It has already begun: Blogs are beginning to embed more and more video, and bloggers are themselves turning into talking heads, thanks to innovations like Bloggingheads.tv. How will science communication fare as this trajectory continues? We don't know, but it's possible that once again, we'll see the isolation of the science community into just one small part of the total space occupied and served by "Television 2.0"—just as we do now with science content on both cable television and in the blogosphere.

In the next chapter and the conclusion, then, we outline a very different strategy for reconnecting science and our society. This approach is not "for profit" and not dependent on developments in new media. Rather, it turns upon mobilizing a new workforce and, further still, a new movement to bring science back in touch with the rest of America. The argument is that we must remake the science-educational "pipeline" to generate more "science ambassadors" who will be adept at engaging in outreach to the rest of our society and culture. At the same time, we must generate the positive energy and enthusiasm needed to inspire pro-science activists who will work tirelessly to bring science to all of the people, in the mode of the ScienceDebate2008 initiative. And this they must do not simply through blogging (though they should hardly neglect it), but through dedicated outreach to the worlds of politics, media, entertainment, and the religious community—as well as to the American public in its very broadest sense.

CHAPTER 10

Is Our Scientists Learning?

I N LATE 2005, THE NATIONAL ACADEMY OF SCIENCES ISSUED ONE OF THE MOST influential studies in its history. Entitled *Rising Above the Gathering Storm*, the report sounded the alarm: The United States, it warned, wasn't producing enough scientists and engineers to keep us competitive for the long haul. Or as the study's authors put it: "The scientific and technological building blocks critical to our economic leadership are eroding at a time when many other nations are gathering strength." Looking worriedly at the emerging science superpowers China and India in particular, the NAS committee, comprising top scientists, university presidents, and industry leaders, called for a large increase in the "number and proportion" of U.S. students who earn science and engineering degrees. The committee also recommended dramatically bulking up K–12 science education.

The American scientific community has long ranked educational change as a top priority, but perhaps never before had it so effectively paired that message with one about the future of the U.S. economy. In a bipartisan way, Congress lurched into action, delivering the America COMPETES Act of 2007 (short for "America Creating Opportunities to Meaningfully Promote Excellence in Technology, Education, and Science"), designed to reinvest in research and science education at all levels. Since then, there have been ongoing struggles to ensure that COMPETES gains adequate funding, an important push and one that continued through the economic stimulus battle of early 2009, which resulted in large and very welcome increases for the National Science Foundation, Department of Energy, National Institutes of Health, and other science-related agencies.

We take no issue with the scientific community's focus on shoring up our economic competitiveness or stoking the engines of innovation. Producing a more scientifically literate workforce can only benefit this country, and elementary and high school science education does need vast improvement. Yet it will be apparent that the *Gathering Storm–* inspired reforms largely neglect the kinds of problems discussed in this book. After all, America doesn't merely need non-scientists to better understand the details of science, or the nature of the scientific method; we need them to see why science matters to their lives and their careers, whether they're working in politics, the media, the corporate world, or some other sector. And we don't merely need the United States to produce more scientists: We need it to produce scientists who have a better understanding of other disciplines, and who are trained in (and value) outreach to the rest of society.

It's certainly wise to keep our eyes on the rearview mirror, and the competitors—most centrally, China—that appear to be coming up fast. But let's not forget that the U.S. science establishment remains the envy of the world. We're producing more Ph.D.s in science and engineering each year. We spend more than any other nation in the biomedical research arena, and in total government-funded research and development. We employ the most scientists, are the chief source of valuable new patents, and publish vastly more peer-reviewed research than all of our competitors (and nearly four times as much as our nearest rival, Japan). None of these facts, however, have helped science garner more serious media attention, or better treatment in Hollywood, or a fairer shake in churches. None of them help bridge the science-society gap.

That's why we must broaden our conception of science education and even, perhaps, of "competitiveness." In truth, there may be as much wrong with the high-level education of scientists as there is with the high school science education of the public. Simply producing more scientists won't solve our cultural problems—or at least, not if we produce them in the same way that we have always done.

The *Gathering Storm* reforms seek to repair the nation's so-called science-education pipeline, which sucks in young students and spits them out at the other end as minted scientists. Yet a survey of this human assembly train shows that those reforms are, at best, only a partial salve for its many miles of corrosion and poor engineering.

The science pipeline's gaping maw is primary school, where far too many students never grasp what science actually means as they zoom in on equations and formulas. They memorize quite a lot of science (Periodic Table of the Elements!), but do their studies resonate for them? Do they see how science will transform the future world they will inhabit? Not according to the National Academy of Sciences, whose 2005 study *America's Lab Report* laments: "Neither . . . scientific literacy—nor an appreciation for how science has shaped the society and culture is being cultivated during the high school years." Too many high school students instead wade through chemistry and physics with their heads down, uninspired by their teachers, never seeing that science's most profound implications reach far beyond the facts they must recite and the chore that is homework.

The media understand this. When the producers of ABC's *The Wonder Years* wanted to cast Kevin Arnold's science teacher, Mr. Cantwell, for instance, they called on Ben Stein. Stein's own words, explaining why he got the part, say it all:

> I am not an actor. I got called by ABC to play the same kind of person I have always been—a big, monotoned nerd, exactly what I was in the movie *Ferris Bueller's Day Off* and what I am in real life when I teach at Pepperdine or UCLA or give expert testimony in securities law cases. I was drafted to play the kids' slightly scary, extremely pessimistic science teacher, Mr. Cantwell, whose slide shows and lectures have often paralleled what was going on in Kevin's life.

Such depictions make it plain why scientific luminaries aren't today's popular role models. For young people who have never met a real scientist, *Hannah Montana*'s Miley Cyrus probably seems a lot more relevant to their lives, and she gets to go on tour.

So, our high schools turn a lot of smart people off science—smart people who instead go on to study law, finance, or business. Yet many students do pursue scientific degrees at the university level, where they embark on a rigorous curriculum that sets them on course to develop a highly rarefied set of skills. They brave the introductory courses deliberately engineered to weed out those lacking the "drive," and persevere through hours of lab work. Many complete honors theses, publish their own research, and come to know the Krebs cycle better than most graduate students.

By the time they obtain their bachelor's degrees, however, many aspiring scientists will already have noticed something troubling. As they toil away, their friends are finding employment in fields like law, sales, and marketing, with high salaries in big cities. In contrast, their certificate can begin to feel much like its title: B.S.

Here in the pipeline is where leakage, or attrition, starts to occur. According to a 2007 study by the Urban Institute, within two years of graduation, 20 percent of students who earn bachelor's degrees in science and engineering have remained in school, but no longer find themselves in a science or engineering field. Another 45 percent have gone out into the world and found jobs, but they're not science- or engineering-related. Who are these lost scientists? Possibly the students who aren't sure they can carry on when the future promises more long hours, teaching and research responsibilities, and an annual stipend as low as $12,000–$35,000 in graduate school.

More attrition occurs between the receipt of a graduate degree in science and the decision to enter a Ph.D. program: 7 percent of students who obtain a master's degree in a scientific field subsequently move into a different academic area, and 31 percent move into the job market in a non-science or non-engineering position. And given that the median time spent getting a Ph.D. is 7.9 years and the median age at the time of doctorate receipt in a science and engineering field is 32.7 years, who can blame them? For those supporting young families, caring for ailing parents, and carrying heavy student loans, pressing on can become a heavy financial burden.

And yet despite such realities, American universities are currently awarding a record number of science and engineering Ph.D.s. From 2002 to 2007, the number of doctorates in these fields grew from 24,608 to 31,801, representing five straight years of increases. These young experts leave their institutions with a firm grasp of experimental protocols and a devotion to statistical rigor. They are fluent in the passive-voiced language of the science world, one that generally lacks metaphors and adjectives and is wholly distinct from the writing style of those trained in the humanities. They are now fully minted *scientists*, and what they have been through to get there—intellectual rewards notwithstanding—has left them worlds apart from those friends who long since departed to work on Wall Street, in Hollywood, on Capitol Hill.

But wait: It's not actually over yet. Those scientists who earn Ph.D.s and are intent on a professorship face a sometimes multiyear probationary period. There are some 48,000 such "postdocs," or post-doctoral fellows, spread across the United States, according to the National Post-doctoral Association. They spend an average of 1.9 years in this very challenging role, which entails the further immersion in research in the interest of someday earning a true faculty job. Fifty-eight percent of postdocs are between thirty and thirty-five years old—right around the age when many people begin families—and 69 percent report being married, with 34 percent having children. Yet average salary is low, in the neighborhood of $40,000 per year; work hours are long, and there is tremendous pressure to publish—or perish, as the saying goes. And then the faculty search begins.

It, too, is brutal. Between 1972 and 2003, the percentage of recent Ph.D.s attaining full-time faculty posts declined dramatically, from 74 to 44 percent—even as, of course, postdoc numbers rose. According to pre-recession figures, the chance of a Ph.D. recipient under age thirty-five winning a tenure-track job has tumbled to only 7 percent. In other words, we're producing more Ph.D. science graduates than ever, yet the traditional academic trajectory affords fewer and fewer job opportunities.

So who remains at the end of pipeline, gets the academic science jobs, runs the labs, and teaches the courses? Those scientists who make it will be the smartest, most motivated, and best equipped, and will probably have benefited from a bit of luck along the way. They will also have developed perspectives and worldviews based on life experiences, sacrifices, and decisions that are vastly distinct from those of the public at large.

The tribulations of the science pipeline explain why today's youngest scientists—who hold in their hands our future, for innovation and for the place of science in American society—might not look fondly on talk of dramatically ramping up the production of U.S. researchers. These young minds are at a time in their careers when Einstein was having his miracle year, and Darwin was aboard the *Beagle*. Yet today they see a dwindling number of academic jobs, and vast numbers of their fellow aspiring scientists in postdoc holding patterns. To quote a painfully eloquent recent blog commenter:

> I'm a recent PhD graduate (Aug' 2008). I'm unemployed. I am valued at negative $75,000 as a result of my school loans. For an increasing number of PhD graduates, there is NO job at the end of the PhD tunnel, unless you opt for the path of the underpaid, undervalued limbo lifestyle of a postdoc. After seeing what my predecessors have suffered on that path (~10 years of postdocing, and STILL no tenure-track job?), I chose NOT to follow in their weary footsteps. I have found that I'm not only overqualified for many positions that I would be happy to hold, but I am also considered by recruiters to be very narrowly-qualified (despite my multidisciplinary interests and skills) for anything at all except being a lab monkey and working for $30,000 a year. Had I to do it over again, I would not choose a PhD, at least not a general science degree. I would have gone to medical or law school, or

perhaps a PhD in public health (a very rapidly growing field). At least after training in these programs, your skill set is clearly defined, and you can be confident that you will have a job post-graduation.

This commentator's frustration underscores a troubling reality: Even as university opportunities are dwindling, the excess in the non-academic job market for scientists and engineers once again appears to lie on the supply side. Every year, according to the Urban Institute, we produce more than three times as many four-year college science and engineering graduates as there are corresponding science and engineering job openings.

Nor does American graduate science education adequately prepare students for scientific jobs outside of the academy—or at least, not if it only provides technical scientific expertise. Bill Bates, vice-president for governmental affairs at the Council on Competitiveness, explains that the United States can probably never hope to produce as many total engineers and scientists as India and China, given their vastly larger populations. Our advantage, then, will lie in producing scientists with "soft skills," such as in writing, speaking, and task management, which are what companies really want.

And yet instead of realigning scientist education to produce more such talents, we're sending today's young researchers into situations of ridiculous competition for traditional positions and grants, far beyond what's healthy to ensure excellence. Our current model does not allow many of our nation's most gifted and dedicated minds to fulfill their potential to become tomorrow's leaders. It trains vast numbers of them in scientific and technical skills, but all too seldom ensures that they gain adequate interdisciplinary, communication, writing, and speaking experience—abilities that not only would help them in the non-academic science job market, but would also make them far better ambassadors, for science, to the rest of society.

So isn't the solution obvious? On the one hand, we need to relieve pressure on the scientific pipeline, create more opportunities for younger scientists, liberate postdocs from holding patterns, and train newly minted scientists to better compete in an uncertain job market. On the other hand, we need to encourage the scientific community to engage in more outreach and produce scientists who are more interdisciplinary and savvy about politics, culture, and the media.

These goals ought to be one and the same. Why not change the paradigm and arm graduate-level science students with the skills to communicate the value of what science does and to get into better touch with our culture—while pointing out in passing that having more diverse skills can only help them navigate today's job market, and may even be the real key to preserving U.S. competitiveness?

Meanwhile, let's encourage public policy makers, leaders of the scientific community, and philanthropists who care about the role of science in our society to create a new range of nonprofit, public-interest fellowships and job positions whose express purpose is to connect science with other sectors of society.

How does all of this square with oft-expressed concerns about the United States falling behind in science? If anything, it's more deeply germane to these concerns than the "more is better" clichés that we often hear. Particularly if we, as a nation, are going to be training even more scientists than in the past, then we've also got to create more interdisciplinary opportunities for them and nurture broader sets of abilities among them. Mostly, this will make them better, more productive scientific workers; but some of these young interdisciplinary scientists should be tapped as "ambassadors" to our broader society. It won't take that many: As Carl Sagan's example shows, when it comes to the media, a single individual can have a dramatic impact. So if we're increasing our scientist ranks anyway, surely there's ample room for training a cadre of communication and outreach experts, and creating subsequent public-interest-oriented jobs for them.

Young scientists already have the minds—and can easily develop the skills—that will allow them to succeed at bridging the science-society gap. The problem is that they rarely get the training. Yes, they receive intensive instruction in their respective fields—but do they learn to tell stories about what they do through narrative, and thereby appeal to the interests of audiences much broader than their group of lab mates? Can they speak in sound bites when necessary? Do they understand the different needs of politicians, journalists, and entertainers for scientific information, and are they prepared to convey their knowledge in that appropriate form?

To create such Renaissance scientists, we must fundamentally change the way we think and talk about science education—and that means rethinking *scientist* education as much as education for high school students or college undergraduates.

One crucial aspect of the endeavor will be making sure that the current trend toward university-level interdisciplinary education not only continues, but comes to encompass far more real risk-taking than it currently does. Graduate programs have been touting their "interdisciplinarity" for decades, and yet in truth, these bridge-building attempts still meet with considerable entrenched resistance. Merely straddling the line between physics and chemistry can pose risks for a student contemplating a career in academia—the traditionalist physicists won't get it on their side, and the traditionalist chemists won't on theirs, either—so just imagine how science communication and media initiatives are likely to fare in most university science departments. To truly set change in motion, we need a dramatic increase in funding for initiatives like the National Science Foundation's IGERT (Integrative Graduate and Education and Research Traineeship) program, a decade-old interdisciplinary endeavor that is clearly ready to be taken to the next level.

If American science truly fears for its competitiveness in the global marketplace, it ought to be expanding and reinventing itself to incorporate new opportunities for young American scientists. The scientist who can write, or design a Web site, or understand patent law, or speak Spanish will be better equipped to face the competition than a scientist who only knows his or her discipline—not to mention a better science communicator. And in the context of the science-education pipeline, these alternative valves will alleviate pressure by opening new pathways for pent-up scientific talent to filter out into society.

Pace the National Academy of Sciences, then, more is not necessarily better. We wouldn't go so far as to suggest—as some have—that producing more scientists is unethical in light of current job prospects; it's too difficult to try to time what the market needs. Rather, we'll take a stand on this point: What America requires isn't necessarily *more* total scientists—or at least, not in isolation. Rather, we need more *well-rounded* scientists.

So as we embark on a course of training more U.S. scientists, let's also seek to ensure that they learn more about politics and the media, that they develop communication skills, and that some proportion of them will leave their universities ready to serve as culture-crossers who engage in outreach to the rest of our society. And let's create jobs, positions, and incentives that will encourage them to do so—stimulating not only scientific innovation, but scientific outreach as well.

The fact is, we don't merely need a smarter population that can regurgitate what's in the textbooks. We need one that cares about science, has it on the radar, sees it as salient and relevant. And we don't simply need a bigger scientific workforce: We need a more cultured one, capable of bridging the divides that have led to science's declining influence.

CONCLUSION

A New Mission for American Science

IN SEPTEMBER 2008, THE MULTIBILLION-DOLLAR LARGE HADRON COLLIDER (LHC), the world's most powerful particle accelerator, started up the test runs that will eventually allow it to smash together protons at an unprecedented speed. The resulting data may ultimately help scientists achieve a deeper understanding of the nature of matter and the universe, and perhaps allow them to discover the mysterious Higgs boson (sometimes dubbed the "God particle"), thereby determining whether particle physics' current "standard model" is correct.

The collider represents a major step forward in the march of science: The profundity of the questions it could answer is hard to overstate. We're talking, literally, about trying to glimpse the true nature of reality. So the fact that the quest is not happening in the United States is deeply symbolic—and saddening. The LHC is instead a project of CERN, the European Organization for Nuclear Research, and funded by European governments. In contrast, Congress killed funding for the nation's planned Superconducting Super Collider in 1993, amid budgetary concerns and a growing sense that with the collapse of the Soviet Union, the need to invest in such massive projects had declined.

That the LHC project is happening abroad, however, has not pre-vented it from becoming the source of a great deal of paranoid pseudo-scientific silliness. Although repeatedly dismissed by physicists, concerns abound that the collider will somehow create mini black holes that will grow to envelop us all or generate "strangelet" particles that transform everything else into their particular form of nastiness. Lawsuits have been filed to stop its operation; CERN researchers have received death threats. And earlier this year, Hollywood released the film version of *Angels & Demons,* the 2000 novel by *The Da Vinci Code* author Dan Brown whose plot turns on villains trying to use antimatter taken from CERN to destroy the Vatican. CERN has endeavored to set the record straight, but the likelihood its Web page will equal the Ron Howard–Tom Hanks movie in viewership is about the same as the probability of the LHC sucking us all into a vacuum.

Is this the future of science we want to see? Do we want to see re-search and knowledge advance, but have a public that does not neces-sarily follow or come along, and at many moments in fact actively resists, or reacts with shock and alarm? As Carl Sagan put it in *The Demon-Haunted World:*

> We've arranged a global civilization in which most crucial elements profoundly depend on science and technology. We have also arranged things so that almost no one understands science and technology. This is a prescription for disaster. We might get away with it for a while, but sooner or later this combustible mixture of ignorance and power is going to blow up in our faces.

At present we're marching steadily toward that outcome. The num-ber of burgeoning scientific fields in which new knowledge threatens to trigger a backlash, because of its potential to transform our societies and our world, is astounding. In the next few decades, pharmaceutical companies may begin producing pills that significantly lengthen the human life span, providing many benefits but also possibly disrupting a key foundation of society—the relationship between the young and

the old. Biotech firms will likely develop synthetic life in the laboratory (cue up the *Frankenstein* movie, please), microscopic organisms
characterized by the minimal amount of DNA necessary to make
them functional, and perhaps engineered to serve particular environmental purposes.

And that's just the beginning. New developments in neuroscience
may someday make possible, on a scientific basis, synthetic "telepathy."
If you can capture a brain state well enough to digitize its informational
content in a computer, then what's to stop you from analyzing its
"thoughts" or even beaming that information to another brain? The
military is already interested. Although scientists themselves protest
that such things are a long way off—and warn against hype, a caution
we should take very seriously—we also know that science constantly
surprises us. If neuroscience doesn't bring about the shock, some other
field will.

Meanwhile, as global warming worsens, many scientists have begun
weighing the prospects of "geoengineering": artificially turning down
the planetary thermostat. The means of doing so already lie within our
grasp—for instance, a geoengineering scheme aiming to mimic a volcanic eruption (known to cool down the earth) by infusing the strato-
sphere with sulfur dioxide particles could probably be implemented
next year. Of course, no one can predict what the full ramifications
would be: Such an intervention would definitely decrease global temperatures, but what other effects would it have? It could change rainfall
patterns, deplete the ozone layer, and who knows what else. Active geoengineering would likely result in mass protests in the streets, if not
open war between countries that would prefer a balmier global temper-
ature (think Russia) and those who stand to fry, like the United States.
But will that stop some government from trying it?

The above represents a very small sample of a much longer list of potential challenges and opportunities that science and technology will pose
for us in the decades to come. Incomprehensibly big changes are coming,
and we need a strong rapprochement between science and our society
before the next high-profile crack-up occurs. We must systematically

reconnect the world of science with the rest of America, and especially its most influential sectors. We need much more respect, and much more dialogue, between separate spheres, so that when the next spate of controversies arrive, they don't threaten to tear us apart.

Most of all, we need science to reestablish its core relevance to American life, to enjoy the standing and visibility it had in the late 1950s and early 1960s (with full accommodation of the lessons learned since then). Otherwise, we'll simply repeat the cycle of ongoing scientific research that few people understand, interrupted by occasional public shock and outrage, and then followed in turn by more societal slumber once everyone forgets again—the politicians most of all—what the scientists are up to.

One can scarcely doubt that the causes of the disconnects we've highlighted are diverse. But that doesn't mean those of us who lament them—those of us who are either the scientific community's allies or its actual members—can be satisfied to lay blame elsewhere without taking action of our own. We must all rally toward a single goal: Without sacrificing the growth of knowledge or scientific innovation, we must invest in a sweeping project to make science relevant to the whole of America's citizenry. We recognize there are many heroes out there already toiling toward this end and launching promising initiatives, ranging from the Year of Science to the World Science Festival to ScienceDebate. But what we need—and currently lack—is the systematic acceptance of the idea that these actions are integral parts of the job description of *scientists themselves*. Not just their delegates, or surrogates, in the media or the classrooms.

The kind of communication we need has, ultimately, very little to do with the in-house conversations about science that currently occur all the time on blogs, and in specialized magazines that serve the scientific community and those who care about it. We have nothing against media directed specifically at those who already nourish an interest in science: We write for such outlets, and we read them. But the past fifty years, and the lessons of previous science-communication efforts, plainly show the danger of simply assuming such information will be enough. Meanwhile, the current crisis of science in the media is the gravest warning yet that unless scientists themselves engage, no one else can be relied upon to do it for them.

At the same time, as we move into the Obama administration, the scientific community must avoid another trap: complacency. Already, we're detecting the sentiment that after eight years of George W. Bush's "war on science," we've finally been saved. There's a real risk that scientists, energized as never before by the abuses of the Bush administration, will lapse into relaxed detachment from the rest of society as Obama begins to govern.

That would be the gravest of errors. First of all, scientists must remain active to ensure that science policy remains a priority in the Obama administration, far beyond a few critical areas, such as energy policy, that our new president has already highlighted. It is still too easy today for science to fall off the radar, and we are far from achieving anything like the changes we need to see to feel truly justified in resting on our laurels. The scientific community has decades of catching up to do when it comes to cementing its place in American society. Maybe we might think about taking a rest if the percentage of Americans subscribing to young-Earth creationism dips below 20—but until then, we must be constantly vigilant.

Let's also avoid the mistaken idea that university-based changes will be all that we need. Yes, universities need to do far more to reward outreach and should train science students to conduct it. But equally important, or perhaps even more so, is inspiring the science grass roots and using the Internet to channel its pent-up energy and enthusiasm, as demonstrated by the ScienceDebate2008 initiative. Science needs tens of thousands of supporters calling for debates during election season, sending e-mail blasts to members of Congress when there's a particular outrage or a bill to support, and raising money through small donations over the Web (as the Obama campaign famously did, and as ScienceDebate has begun to do) to support further outreach and mobilization.

When you contrast how quickly ScienceDebate2008 emerged on the scene with how slowly universities and scientific institutions have sometimes reacted to change or political problems, it's obvious we are entering an entirely new universe for science-centered activism.

Certainly, we have no wish to exonerate politicians, the media, or the entertainment industry for failing to help us understand and appreciate science, or to forgive the conservative religious community for so regularly attacking it. But the fact remains that scientists, and the people who care about their work, know best what is being missed, why it matters, and indeed, how the science-society gap places our entire future at risk. Moreover, they have the talent, the knowledge, and in many cases the resources to turn things around.

So what are we waiting for? It's time to have more vision. We must set the course for how we want to improve our world, rather than simply reacting to the shifting priorities of various political constituencies or administrations. And our mission is simply massive. Long after his "two cultures" lecture, C. P. Snow expressed the nature of change we need succinctly, yet powerfully:

> We require a common culture in which science is an essential component. Otherwise we shall never see the possibilities, either for evil or good.

We agree, and we take this assertion even further. Science is not merely culture's "essential component," and we don't just have to mend the rift between science and culture: We have to create a perfect union. Science itself must become the common culture. Like Sagan, and like Snow, we're certain our future depends on it.

NOTES

Chapter 1

1   *semantic exercise:* For a further, valuable discussion of the meaning of the word *planet* and whether defining it is a scientific matter, see Phil Plait, "Congratulations! It's a Planet!" *Bad Astronomy* blog, August 15, 2006, http://blogs.discovermagazine.com/badastronomy/2006/08/15/congratulations-its-a-planet/.

1   *as a group of scientists:* See Dava Sobel, "Pluto's Brave New Worlds," *Washington Post,* August 16, 2006, http://www.washingtonpost.com/wp-dyn/content/article/2006/08/15/AR2006081501124.html.

1   *the IAU rejected that compromise:* For the official IAU decision, see IAU 2006 General Assembly, Result of the IAU Resolution Votes, August 26, 2006, http://www.iau.org/public_press/news/release/iau0603/.

2   *"No do-overs":* Tim Kreider, "I ♥ Pluto," *New York Times,* August 23, 2006, http://www.nytimes.com/2006/08/23/opinion/23kreider.html.

2   *remarked Alan Stern:* Quoted in Robert Roy Britt, "Scientists Decide Pluto's No Longer a Planet," *MSNBC.com,* August 24, 2006, http://www.msnbc.msn.com/id/14489259/.

3   *over $100 billion annually in federal funding:* The federal science budget—technically, the "research and development," or "R&D" budget—is a complicated creature, but allow us to at least provide some sense of scale. In President Bush's February 2008 budget request for fiscal year 2009, proposed funding for basic and applied research amounted to $57.3 billion in total. However, the larger part of the R&D budget is not for research but for development, and so the total request for both areas combined was $147.4 billion. See American Association for the Advancement of Science, "R&D in the FY 2009 Budget," http://www.aaas.org/spp/rd/fy09.htm.

It's also illuminating to compare the total federal R&D investment to that being undertaken by private industry. In 2007, industry accounted for two-thirds of total U.S. R&D, or roughly double the total federal investment. For understandable reasons, however, industry focuses far more heavily on development and much less on basic research. Again, see AAAS, http://www.aaas.org/spp/rd/guitotal.htm.

3   *a million lives per year:* According to the Centers for Disease Control and Prevention, we've reduced morbidity by 99 percent or more for the following diseases: smallpox, diphtheria, measles, polio, and rubella. Averaged over the course of the twentieth century, these five diseases killed more than 600,000 people annually. In 2006, they killed fewer than 100. Meanwhile, we've also dramatically reduced deaths from the mumps and pertussis, which used to kill upward of 350,000. See Dr. Melinda Wharton, CDC PowerPoint, http://www.cga.ct.gov/coc/PDFs/immunization/wharton_ppt.pdf.

3   *to believe that global warming is real:* See the Pew Research Center, "A Deeper Partisan Divide over Global Warming," May 8, 2008, http://people-press.org/report/417/a-deeper-partisan-divide-over-global-warming.

4   *Just 18 percent . . . and even fewer:* Mary Woolley and Stacie M. Propst, "Public Attitudes About Health-Related Research," *Journal of the American Medical Association*, Vol. 294, No. 11, September 21, 2005.

4   *44 percent of the respondents:* Survey on the "State of Science in America," Museum of Science and Industry, Chicago, March 20, 2008. Results available online at http://www.stateofscience.org. For a more elaborate discussion of findings, see Eric Berger, "Who Are America's Science Role Models? (It's Ugly, Folks)," March 21, 2008, http://blogs.chron.com/sciguy/archives/2008/03/who_are_america.html.

5   *widely denounced for a disdain of science:* Chris Mooney, *The Republican War on Science* (New York: Basic Books, 2005).

5   *his classic 1962 work:* Richard Hofstader, *Anti-Intellectualism in American Life* (New York: Vintage, 1962).

5   *deep-thinking scientist of the first rank:* Joyce E. Chaplin, *The First Scientific American: Benjamin Franklin and the Pursuit of Genius* (New York: Basic Books, 2006).

5   *Alexis de Tocqueville:* For this reading of de Tocqueville, we're indebted to Steven Shapin, *The Scientific Life: A Moral History of a Late Modern Vocation* (Chicago: University of Chicago Press, 2008), p. 43.

5   *A 2008 analysis:* We first noticed these data thanks to Matthew Nisbet's blog, *Framing Science,* March 17, 2008, http://scienceblogs.com/framing -science/2008/03/if_you_watch_five_hours_of_cab.php. The study in question is from the Project for Excellence in Journalism, "The State of the News Media 2008," http://www.stateofthenewsmedia.org/2008/ narrative_cabletv_contentanalysis.php?cat=1&media=7.

6   *weekly science or science-related sections:* Cristine Russell, "Covering Controversial Science: Improving Reporting on Science and Public Policy," 2006 Working Paper, Joan Shorenstein Center on the Press, Politics, and Public Policy, http://www.hks.harvard.edu/presspol/research_publications/ papers/working_papers/2006_4.pdf.

6   *modern vaccine-skeptic movement:* January 12, 2009, interview with journalist Arthur Allen, author of *Vaccine: The Controversial Story of Medicine's Greatest Lifesaver* (New York: W. W. Norton, 2007).

6   *degree from the "University of Google":* People Magazine, "My Autistic Son: A Story of Hope," September 20, 2007, http://www.people.com/ people/article/0,,20057803,00.html.

6   *try on information sources:* We want to thank Matthew Nisbet for playing a key role in making us think about the problem of media "fragmentation," which we discuss further in subsequent chapters. There, we have also cited other authors who have outlined the problem, such as Cass Sunstein (in 2001's *Republic.com*) and Princeton University's Markus Prior.

7   *a growing trend:* Hugh Hart, "An Injection of Hard Science Boosts the Prognosis for TV Shows," *Wired,* December 5, 2008, http://blog.wired .com/underwire/2008/12/science-fact-fa.html.

8   *Obama administration could find its hands tied:* See, for example, Neal Lane and Leslie Berlowitz, "Where to Spend Our Research Dollars," *Science Progress,* January 22, 2009, in which they note, "once we move beyond crisis-response mode, discretionary funding is likely to be severely constrained," http://www.scienceprogress.org/2009/01/where-to-spend-our-research-dollars/.

9   *funding for energy innovation was in steep decline:* Daniel M. Kammen and Gregory F. Nemet, "The Incredible Shrinking Energy R&D Budget," *Issues in Science and Technology* (Fall 2005), http://www.issues.org/ 22.1/index.html.

9   *scientific research has been a core driver:* National Academy of Sciences, *Rising Above the Gathering Storm: Energizing and Employing America for*

a Brighter Economic Future, Committee on Science, Engineering, and Public Policy (Washington: National Academies Press, 2007).

9 *failed to keep pace with inflation:* See American Association for the Advancement of Science, "Final Stimulus Bill Provides $21.5 Billion for Federal R&D," February 16, 2009, http://www.aaas.org/spp/rd/stim09c.htm.

10 *retardation of human aging:* See Robert Butler et al., "New Model of Health Promotion and Disease Prevention for the 21st Century," *British Medical Journal*, July 8, 2008, http://www.bmj.com/cgi/content/extract/337/jul08_3/a399.

10 *guilt or innocence:* See Michael S. Gazzaniga and Megan S. Steven, "Neuroscience and the Law," *Scientific American*, April 15, 2005, http://www.sciam.com/article.cfm?id=neuroscience-and-the-law.

10 *actively manipulate the planet's climate and weather:* See David G. Victor et al., "The Geoengineering Option," *Foreign Affairs* (March–April 2009), http://www.foreignaffairs.org/20090301faessay88206/david-g-victor-m-granger-morgan-jay-apt-john-steinbruner-katharine-ricke/the-geoengineering-option.html.

11 *rewrite the nation's educational curriculum:* See John L. Rudolph, *Scientists in the Classroom: The Cold War Reconstruction of American Science Education* (New York: Palgrave, 2002).

11 *fire rockets from their backyards:* Perhaps ABC's *Good Morning America* put it best: Covering the innovative 2008 World Science Festival in New York City, the program remarked that the event organizers aimed to shift science away from the "chalk-dusted fringes" of American life and back into our "cultural center" where it belongs (but does not currently reside). "Science Rocks! The First World Science Festival in New York Is Hoping to Turn Geek Chic," *Good Morning America,* June 1, 2008.

12 *Snow delivered a famous speech:* C. P. Snow, *The Two Cultures* (Cambridge: Cambridge University Press, 1993).

Chapter 2

13 *"scientific illiteracy":* For the standard definition of scientific literacy, see a summary in Bauer et al., "What Can We Learn from 25 Years of PUS Survey Research? Liberating and Expanding the Agenda," *Public Understanding of Science*, Vol. 16 (2007), pp. 79–95. See also Rüdiger C. Laugksch, "Scientific Literacy: A Conceptual Overview," *Science Educa-*

tion, Vol. 84, No. 1 (December 14, 1999), pp. 71–94. One researcher who has been particularly influential in limning the problem of scientific illiteracy is Jon Miller of Michigan State University. He has proposed a metric called "civic scientific literacy," combining factual and definitional knowledge of "scientific constructs" with an understanding of the "process or nature of scientific inquiry." See, for example, Jon D. Miller, "Scientific Literacy: A Conceptual and Empirical Review," *Daedalus*, Vol. 112, No. 2 (Spring 1983), pp. 29–48, and Miller, "The Measurement of Civic Scientific Literacy," *Public Understanding of Science*, Vol. 7 (1998), pp. 203–223.

13   *can't read the* New York Times *science section:* Jon D. Miller, "Public Understanding of, and Attitudes Toward, Scientific Research: What We Know and What We Need to Know," *Public Understanding of Science,* Vol. 13 (2004), pp. 273–294.

14   *blame is said to lie with "the public":* In a series of lectures delivered jointly with Chris in 2007 and 2008, and in personal conversations, Matthew Nisbet helpfully highlighted many of the problems with the so-called deficit model critiqued in this chapter. We greatly appreciate his role in helping to bring the scholarly literature on this subject to our attention. In addition, in those lectures Nisbet also singled out for criticism the related "popular science model," or the idea that popular-science media outlets can adequately educate Americans about science.

14   *citizens of other nations:* See Jon D. Miller, "The Public Understanding of Science in Europe and the United States," paper presented at the 2007 annual meeting of the American Association for the Advancement of Science, San Francisco, February 16, 2007.

14   *Residents of the European Union:* Ibid.

14   *As Mark Twain put it:* We found this quotation in Paul Offit's excellent book, *Autism's False Prophets: Bad Science, Risky Medicine, and the Search for a Cure* (New York: Columbia University Press, 2008).

15   *Scientific research has soundly refuted:* For the definitive study, see Institute of Medicine, *Immunization Safety Review: Vaccines and Autism* (Washington, DC: National Academies Press, 2004).

15   *Global warming isn't happening:* For the definitive refutations of such nonsense, we recommend the top global warming blog, http://www .realclimate.org.

16   *scholars . . . have largely discarded:* Some of the relevant literature was assigned for a daylong "boot camp" about science communication cotaught

by Matthew Nisbet and Chris at Caltech in summer 2008. For a list of readings, see http://sass.caltech.edu/events/boot_camp.shtml.

16   *the "deficit model":* For further familiarizing us with the problems inherent in deficit thinking, we'd like to thank David Guston and Naomi Oreskes.

17   *"you're an idiot":* In Randy Olson's excellent documentary *Flock of Dodos*, a pro-evolution scientist uses exactly these words in describing how to respond to the "intelligent design" movement: "I think people have to stand up and say, you know, you're an idiot."

17   *"A deficient public cannot be trusted":* Bauer et al., "What Can We Learn from 25 Years of PUS Survey Research?"

18   *the importance of science to politics, policy, and our collective future:* Or as Michigan State's Jon Miller put it in an influential article: "awareness of the impact of science and technology on society and the policy choices that must inevitably emerge." Jon D. Miller, "Scientific Literacy: A Conceptual and Empirical Review," *Daedalus*, Vol. 112, No. 2 (Spring 1983), pp. 29–48.

19   *scientific leaders still enjoy more public confidence:* See National Science Foundation, *Science and Engineering Indicators 2008,* Chap. 7, http://www.nsf.gov/statistics/seind08/c7/c7h.htm.

19   *Science's ranking vis-à-vis other news topics:* Ibid.

19   *highly superficial degree of appreciation:* To make sense of the public's seemingly contradictory sentiments about science, we like a recent comment by University of Chicago physicist Michael Turner, who remarked that the appreciation of science in this country is "a mile wide and a nanometer thick." Aspen Institute, "Science and the Public Sphere," 2007, http://fora.tv/2007/07/03/Science_and_the_Public_Sphere.

21   *superior or inferior, smarter or dumber:* To be sure, in Snow's original formulation, he bore down considerably harder on the literary types, remarking at one point that whereas scientists have "the future in their bones," the literati respond by "wishing the future did not exist." But at base he painted a deeply resonant picture of two groups of very smart but atomized people, pulled apart by vastly different assumptions.

21   *Now things have flipped:* By 1989, the distinguished scholar of Victorian literature Gillian Beer could remark that although "salient perhaps at the time," Snow's complaints "seem to treat remotely of a dwindling class of literateurs, *not* our main problem now." See Gillian Beer, "Translation or Transformation? The Relations of Literature and Science," *Notes and*

Records of the Royal Society of London, Vol. 44, No. 1 (January 1990), pp. 81–99.

21   *has become conspicuously less influential:* Since Snow's time, we have also seen many other academic changes, including a great expansion of the disciplines (and increased specialization within them) and a huge growth of the social sciences. For more on academic change over the past several decades, see Stefan Collini's introduction to *The Two Cultures* (Cambridge: Cambridge University Press, 1993).

21   *not one but many "cultures":* One might perhaps add other "cultures" as well to this analysis; for instance, scientific culture and legal culture don't mesh particularly well. Although our analysis is not completely exhaustive, then, we believe our focus on political, media, entertainment, and religious culture does go a long way toward capturing the most important disconnects.

Chapter 3

25   *"exhibited as lions":* Quoted in Daniel Greenberg, *The Politics of Pure Science: An Inquiry into the Relationship Between Science and Government in the United States* (1967; New York: Plume Books, 1971 [paperback]), p. 95.

25   *swept up in the national mission:* As the science historian Steven Shapin observes, "During the war itself, mobilized scientists were generally too busy to reflect on what was happening, and, if they were not too busy, security considerations prevented any such public reflections. Subsequently, they struggled to make sense of their experiences. Some felt badly about what they had done; others said they experienced no guilt whatever. But all American scientists now enjoyed the fruits of wartime military labors in the form of vastly increased governmental and industrial funds; enhanced access to what C.P. Snow came to call the 'corridors of power'; a hugely expanded job-market for academic, industrial, and government scientists; and heightened public respect for scientists' power. Scientists had never before possessed such authority, largesse, civic responsibility, and obligations. By free choice or not, some scientists now lived the *vita activa*, and, while there were still consequential worries about the extent to which they were indeed 'normal citizens,' they had never been more integrated into the civic sphere." *The Scientific Life: A Moral History of a Late Modern Vocation* (Chicago: University of Chicago Press, 2008), p. 65.

25 *"Science offers a largely unexplored hinterland":* See *Science: The Endless Frontier,* A Report to the President by Vannevar Bush, Director of the Office of Scientific Research and Development, July 1945, http://www.nsf.gov/about/history/vbush1945.htm. Correspondence between Roosevelt and Bush included.

26 *Independent entrepreneurs:* See Bruce L. R. Smith, *American Science Policy Since World War II* (Washington, DC: Brookings Institution, 1990).

26 *European scientists:* Greenberg, *The Politics of Pure Science,* Chap. 3, "When Science Was an Orphan."

26 *14 percent per year:* This figure is in constant dollars. See Smith, *American Science Policy Since World War II,* p. 39.

26 *"a disastrous blow":* Quoted in Marc Pearl, "On the Offense for Science and Technology Education," *Science Progress,* October 4, 2007, http://www.scienceprogress.org/2007/10/on-the-offense-for-science-and-technology-education/.

27 *"they will have won the solar system":* Quoted in Keay Davidson, *Carl Sagan: A Life* (New York: John Wiley and Sons, 1999 [paperback]), p. 114.

27 *approached $500 million:* National Science Foundation, "A Brief History," July 1994, http://www.nsf.gov/about/history/nsf50/nsf8816.jsp.

27 *grown to $12.2* billion*:* Greenberg, *The Politics of Pure Science,* p. 8.

27 *science-intensive curricula:* For the history of scientists' makeover of the educational curriculum, see John L. Rudolph, *Scientists in the Classroom: The Cold War Reconstruction of American Science Education* (New York: Palgrave, 2002).

27 *swept into the public schools:* This was when the teaching of evolution became prominent in American public education, setting the stage for many later battles.

27 *"This bunch of scientists":* Quoted in John S. Rigden, "Eisenhower, Scientists, and Sputnik," *Physics Today,* June 2007, http://ptonline.aip.org/journals/doc/PHTOAD-ft/vol_60/iss_6/47_1.shtml.

28 *"the most important occupational group":* C. P. Snow, "The Moral Un-Neutrality of Science," lecture delivered in 1960 and published in *Science,* Vol. 133, No. 3448, pp. 255–262, January 27, 1961.

28 *nonmilitary science funding fell:* See American Association for the Advancement of Science, Trends in Nondefense R&D by Function, FY 1953–2009, based on Budget of the U.S. Government FY 2009. Our thanks go to Kei Koizumi for providing this information.

28 *the prominence of the scientific elite in advising our leaders:* As science journalist Daniel Greenberg remarked in 1970, though scientists may have
been able to "ride a crest of Cold War concern," more recently there had
been a "striking . . . decline of the scientific community's influence in
Washington." *The Politics of Pure Science*, pp. xviii–xix.

29 *new mood of "questioning authority":* One book contributed massively to the
questioning of science in this period: science historian and philosopher
Thomas Kuhn's 1962 work, *The Structure of Scientific Revolutions* (Chicago:
University of Chicago Press). Today, when Kuhn's insights have been thoroughly worked over and make sense to many working scientists, it's hard to
read the book as a fundamental assault upon the scientific process or way of
knowing. But it certainly did undermine some naive claims about science's
strict objectivity, such as the always dubious notion that scientists are mere
fact machines rather than human beings with wants, beliefs, personal commitments, pet ideas. Surveying the history of science, Kuhn found that the
growth of knowledge doesn't seem to proceed in a linear fashion, through
the steady accumulation of facts about the world. Rather, it occurs in sudden "revolutions" or "paradigm shifts," in which one scientist or group of
scientists breaks away from an old regime or way of thinking, unable to accept it any longer. In the process, the renegades often face powerful resistance from defenders of the older "paradigm"—defenders who themselves
are often the older scientists in sheer years. Kuhn's argument, bringing out
the role of more subjective factors within the scientific process, seemed to
help knock science off its pedestal.

29 *status of the presidential science adviser:* For the history of presidential science advising see Gregg Herken, *Cardinal Choices: Presidential Science
Advising from the Atomic Bomb to SDI* (Stanford, CA: Stanford University Press, 2000).

29 *Nixon fired his science advisers outright:* Institutionalized presidential science advice would return under President Gerald Ford, but more controversies ensued in the Reagan administration, when adviser George
Keyworth came under intense criticism from scientists after staunchly
supporting Reagan's Star Wars program.

30 *"What the pure scientist basically wants":* Quoted in Greenberg, *The Politics of Pure Science*, p. xvii.

30 *"Modern science not only is inscrutable":* Ibid., p. 35.

30 *scientific journals numbered:* Figures from B. V. Lewenstein, "Science
Books Since World War II," in D. P. Nord, M. Schudson, and J. Rubin,

eds., *The Enduring Book: Publishing in Post-War America* (Chapel Hill: University of North Carolina Press, in press 2009).

31   *"outside of our own competence":* Stephen Jay Gould, "Take Another Look," *Science,* Vol. 286, No. 5441, p. 899, October 29, 1999.

31   *came into question:* By 1990, Brookings Institution scholar Bruce L. R. Smith could describe the change that had occurred since the post-Sputnik days thusly: "Society's support for science had been based on the assumption that progress in the various scientific disciplines would ultimately lay the foundation for a better life for all Americans. Social improvements of all kinds would follow when the nation's collective intelligence was brought to bear on the most pressing problems. But as Americans lost confidence in this premise, as their optimism about the future became tinged with pessimism, the foundations of society's support for science—and scientists' faith in themselves—eroded." Smith, *American Science Policy Since World War II,* p. 77.

31   *scientists did relatively little to counter the trend:* Not that it went unnoticed. In a 1971 speech, William Bevan, a psychologist then serving as executive officer of the American Association for the Advancement of Science, remarked to his colleagues, "Our emphasis on excellence in individual performance has fostered a psychology of elitism that has made both our enterprise and our body of knowledge esoteric and increasingly inaccessible to the layman at all levels of society." Quoted in Sally Gregory Kohlstedt, Michael M. Sokal, and Bruce V. Lewenstein, *The Establishment of Science in America: 150 Years of the American Association for the Advancement of Science* (New Brunswick: Rutgers University Press, 1999), pp. 139–140.

31   *some attempts to reach out:* In 1951, for instance, the American Association for the Advancement of Science designed an ambitious plan to tackle "the broader external problem of the relation of science to society," and thereby make science "better understood by government officials, by businessmen, and indeed by all the people." Scientists felt in a very evangelical mood in those days; but over the next several decades, the AAAS would struggle with how to pursue its outreach goals, amid internal clashes over strategy. See Bruce Lewenstein, "The Meaning of 'Public Understanding of Science' in the United States After World War II," *Public Understanding of Science,* Vol. 1 (1992), pp. 45–68.

31   *"we said quite the other thing":* Quoted in Jim Hartz and Rick Chappell, *Worlds Apart: How the Distance Between Science and Journalism Threat-*

ens America's Future, 1997, First Amendment Center, Freedom Forum, p. 9.

32 *fairly flagrant "deficit" assumptions:* Given the chance to rewrite physics education, for example, the scientific elite from MIT and elsewhere more or less banished from the curriculum the standard technological applications and examples that made physics relevant to everyday life. They didn't want students confusing "basic science" with technology; rather, they wanted to teach the most rigorous, cutting-edge science: The science they themselves were doing. The hard stuff. They aimed for nothing less than to use the schools to produce "the rational man," meaning citizens who would think about science much in the way scientists themselves did. It was a noble—and yet perhaps not exactly realistic—endeavor. For more details, see John L. Rudolph, *Scientists in the Classroom: The Cold War Reconstruction of American Science Education* (New York: Palgrave, 2002).

32 *students' science test scores remained "stagnant":* National Academy of Sciences, *America's Lab Report: Investigations into High School Science* (Washington, DC: National Academies Press, 2005), quote from p. 1, http://www.nap.edu/catalog.php?record_id=11311.

32 *In one contemporary survey:* See Lewenstein, "The Meaning of 'Public Understanding of Science.'"

32 *It was never clear how popular "popular science" could actually become:* For a more detailed overview of these efforts and their problematic nature, see ibid.

32 *"preaching to the converted":* Ibid.

33 *Carl Sagan:* For the details of Sagan's life related in these pages we have drawn upon two biographies, Keay Davidson, *Carl Sagan: A Life* (New York: John Wiley and Sons, 1999 [paperback]), and William Poundstone, *Carl Sagan: A Life in the Cosmos* (New York: Henry Holt, 1999). Further specific references follow.

33 *the true goad toward celebrity:* Frederic Golden, "The Cosmic Explainer," *Time*, October 20, 1980.

33 *"the greatest media work in popular science of all time":* Quoted in Davidson, *Carl Sagan: A Life*, p. 378.

33 *an estimated 500 million viewers:* The figure is from ibid., p. xiv.

33 *roughly twice per year over the span of more than a decade:* The figure is from ibid., p. 264. Poundstone, *Carl Sagan: A Life in the Cosmos*, agrees, p. 261.

34   *grossed $170 million at the box office: Contact* earnings figures from Box Office Mojo, http://boxofficemojo.com/movies/?id=contact.htm.

34   *resisted academic specialization:* Davidson writes, "Sagan was the multidisciplinary scholar par excellence, the Renaissance man so uncommon in the age of specialization, of industrialized academia, where the divisions of labor are as real as in Henry Ford's factories." *Carl Sagan: A Life,* p. 41.

34   *imply for our religious faith:* On this last and most explosive of points, Sagan took a wise line. He remained personally skeptical of religious belief and subjected the arguments for the existence of God to strong scrutiny, finding them wanting (see Carl Sagan, *The Varieties of Scientific Experience: A Personal View of the Search for God,* ed. Ann Druyan [New York: Penguin, 2006]). But at the same time he managed to be both inoffensive and tolerant. In 1984, he traveled to the Vatican to meet with Pope John Paul II and discuss the threat of "nuclear winter," Sagan's theory about the possibly apocalyptic consequences of nuclear war that so enraged the Reagan administration. Especially late in life, Sagan the unbeliever realized there was much to be gained by stressing the elements that science and religion have in common and arguing that the two can work together—the only stance, ultimately, that will help science integrate itself into a pluralistic society like our own. See also Matthew Nisbet, "Sagan: Framing Shared Values Between Science and Religion," September 18, 2007, http://scienceblogs.com/framing-science/2007/09/carl_sagan_on_framing_the_shar.php.

34   *"the awe of science overcame the indifference to it":* Frederic Golden, "The Cosmic Explainer," *Time,* October 20, 1980.

34   *a rejuvenated popular-science movement:* Bruce V. Lewenstein, "Was There Really a Popular Science 'Boom'?" *Science, Technology and Human Values,* Vol. 12, No. 2 (Spring 1987), pp. 29–41. The next several paragraphs are indebted to Lewenstein's insightful research on the popular science "boom," and its later "bust."

34   *"the general interest magazines of a new age":* Quoted in ibid.

35   *leaders of the scientific community also deserved a share of the blame:* Allegedly, they had failed to support the popular-science initiatives of the late 1970s and early 1980s. In fact, these were in large part initiated by journalists, not sparked by scientists or their institutions. See ibid.

35   *"more important things to do":* Quoted in ibid.

35   *a nationwide survey of researchers:* Sharon Dunwoody and Michael Ryan, "Scientific Barriers to the Popularization of Science in the Mass Media,"

Journal of Communication, Vol. 35, No. 1 (Winter 1985), pp. 26–42. The interviewees also agreed that standard scientific training did not adequately prepare scientific acolytes to engage with the media.

35   *a "societal cockfight":* Daniel Yankelovich, "Changing Public Attitudes to Science and the Quality of Life: Edited Excerpts from a Seminar," *Science, Technology, and Human Values*, Vol. 7, No. 39 (Spring 1982), pp. 23–29. Americans, explained Yankelovich, were losing faith in the proposition that had made science so central to their identities following World War II—that it would solve society's problems, generate economic growth, spur progress through technology. Just 52 percent of the public now believed this, according to Yankelovich, and the doctrine was much more trusted by the old than the young. At the same time, he added, "cultural changes born in the 1960s and 1970s" were provoking a "fierce reaction" from the more conservative elements of society, which were now organized into groups like Jerry Falwell's Moral Majority. Highlighting the budding fight over the teaching of evolution—which would reach the Supreme Court in 1987—Yankelovich warned that "science and technology will not be permitted to stand aloof from this values controversy. On the contrary, they will be plunged into the middle of it."

35   *worrying loudly about the gap:* Daniel Yankelovich, "Science and the Public Process: Why the Gap Must Close," *Issues in Science and Technology* (Fall 1984), pp. 6–12. The science community, Yankelovich argued, had brokered a "social contract" with the American public and policy makers that garnered for it "creative isolation" to pursue research—surely a very advantageous and desirable arrangement. Yet this had led to a situation in which the technological sophistication of science was coming to vastly outstrip the public's ability to grapple with complex problems, especially those having a scientific component—issues ranging from acid rain to arms control. Yankelovich therefore called for "upgrading the political literacy of scientists as a prerequisite for two-way communication" with policy makers and the public, and he added that this should happen even if to some extent it impeded "the progress of scientific accomplishment."

36   *"better education in our schools":* Quoted in Lewenstein, "Was There Really a Popular Science 'Boom'?"

36   *the case for retreat:* Leon E. Trachtman, "The Public Understanding of Science Effort: A Critique," *Science, Technology and Human Values*, Vol. 6, No. 36 (Summer 1981), pp. 10–15.

36 *declining scientific credibility with the public:* This is precisely what we see so often today, especially in the coverage of medical and health issues: One tentative epidemiological study says drink coffee (or something similar), another finds the opposite, and people get very angry at science for confusing them and wreaking havoc with their diets.

37 *smiling on creationism:* Reagan had already humored the creationists on the campaign trail, and upon taking office appointed a presidential science adviser—a *Ph.D.*—who did the same before Congress. See Chris Mooney, *The Republican War on Science* (New York: Basic Books, 2006 [paperback]), p. 36.

37 *Reagan's policy . . . went forward anyway:* For a full account, see William Broad, *Teller's War: The Top-Secret Story Behind the Star Wars Deception* (New York: Simon and Schuster, 1992).

37 *an anti-SDI petition:* This episode is related in Davidson, *Carl Sagan: A Life,* p. 358.

38 *target of attacks from William F. Buckley:* See ibid., pp. 371–372.

38 *newly formed conservative think tank:* For more on the history of the George C. Marshall Institute, see Naomi Oreskes and Erik Conway, "Challenging Knowledge: How Climate Science Became a Victim of the Cold War," in Robert N. Proctor and Londa Schiebinger, eds., *Agnotology: The Making and Unmaking of Ignorance* (Palo Alto, CA: Stanford University Press, 2008), pp. 55–89. For a further discussion, see Oreskes and Conway, *Fighting Facts* (New York: Bloomsbury Press, in press 2009).

38 *Nevada nuclear test site:* See Davidson, *Carl Sagan: A Life,* p. 376.

38 *thawing of relations with the Soviet Union:* For Sagan's political influence in the Soviet Union and in the United States, see Poundstone, *Carl Sagan: A Life in the Cosmos,* pp. 318–319.

38 *top outlets for sowing doubts:* Again, see Oreskes and Conway, "Challenging Knowledge: How Climate Science Became a Victim of the Cold War."

39 *1996 Telecommunications Act:* See Common Cause, online report, "The Fallout from the Telecommunications Act of 1996: Unintended Consequences and Lessons Learned," 2005.

39 *a nasty letter:* See Davidson, *Carl Sagan: A Life,* pp. 202–203.

39 *"Quite a few scientists in those days":* Quoted in ibid., p. 203.

39 *History repeated itself in 1992:* For Sagan's spurning by the National Academy of Sciences see ibid., pp. 389–392, and Poundstone, *Carl*

Sagan: A Life in the Cosmos, pp. 355–357. This section draws upon both sources as well as other reports.

39   *distinction in original scientific research:* Scientists can and do differ on whether Sagan's published work was "enough" for NAS membership (a somewhat subjective determination). For a strong argument, based on Sagan's curriculum vitae, that he deserved membership, see Michael Shermer's account in *The Borderlands of Science*, http://www.michael shermer.com/borderlands-of-science/excerpt/.

39   *Sagan's nomination proved divisive:* See Faye Flam, "What Should It Take to Join Science's Most Exclusive Club?" *Science*, Vol. 256, May 15 (1992), who noted: "Members of the academy are sworn to secrecy on the details of how the new candidates are inducted, but insiders made it clear that this was one of several historical cases in which the debate over a nominee grew to involve the entire membership after achieving consensus in a nominee's own discipline."

40   *In their treatment of Sagan:* Although Sagan lost his bid for academy membership, two years later he did receive the National Academies' Public Welfare medal. Still, this stunning slight to the most famous scientist in America sent a clear signal about how some in the scientific community regarded those accused of being mere "popularizers." We want to be clear that our criticism here is not directed at all of science— after all, roughly half of the academy members did vote in Sagan's favor. It's the other half that worries us.

Chapter 4

41   *science books seemed to be breaking through:* Our discussion of the popular-science book publishing phenomenon is indebted to B. V. Lewenstein, "Science Books Since World War II," in D. P. Nord, M. Schudson, and J. Rubin, eds., *The Enduring Book: Publishing in Post-War America* (Chapel Hill: University of North Carolina Press, in press 2009).

41   *paranormalist schlock was filling the airwaves:* See Matthew Nisbet, "Back from Outer Space: With the End of *The X-Files*, the 1990s Infatuation with UFOs May Be Dwindling. Will Psychics Take Their Place?" *American Prospect Online*, May 24, 2002, http://www.prospect.org/cs/articles? article=back_from_outer_space. See also his later comment: "According to the data, cultural fascination with UFOs reached a historic peak in 1996, and remained at an all-time high in 1997. The trend was boosted

by the fiftieth anniversary of the alleged UFO crash at Roswell, New Mexico, and the popularity of entertainment products such as *Independence Day* and *The X-Files*. However, attention to aliens plummeted in 1999, and has yet to recover significantly. . . . Media fascination with psychics follows a similar trend. This topic peaked in popular culture at a historic high in 1999 and 2000, and then sharply declined in 2001." Nisbet, "Cultural Indicators of the Paranormal: Tracking the Media/Belief Nexus," *Skeptical Inquirer Online*, March 22, 2006, http://www.csicop.org/scienceandmedia/indicators/.

42 *None other than CNN's Larry King:* Chris Mooney, "King of the Paranormal," *Skeptical Inquirer Online*, July 31, 2003, http://www.csicop.org/doubtandabout/larryking/.

42 *all these big toys:* For the death of the Superconducting Super Collider and science's forays with the Gingrich budget cutters, see Daniel Greenberg, *Science, Money, and Politics: Political Triumph and Ethical Erosion* (Chicago: University of Chicago Press, 2001).

42 *Newt Gingrich's Republicans:* See Chris Mooney, "Defenseless Against the Dumb," in *The Republican War on Science* (New York: Basic Books, 2005), Chap. 5, for a discussion of the Gingrich Republicans and the dismantling of congressional science advice.

42 *"I have a foreboding":* Carl Sagan, *The Demon-Haunted World: Science as a Candle in the Dark* (New York: Ballantine Books, 1996), p. 25. A similar tone leaped from the literature of one subculture of the scientific community, the skeptic movement, where the sense of alarm over a slew of fads that were deemed irrational—ranging from alternative medicine, to all kinds of New Ageism, to the aforementioned wave of psychics and UFO obsessives—had created a near-siege mentality by the late 1990s. See *Skeptical Inquirer* magazine, whose archives can be read at http://www.csicop.org/si/online.html.

43 *the "only news":* See Peter Catalano, "A Dose of Reality Emerges in LA," *Los Angeles Times,* November 28, 1991.

43 *a "third culture":* The phrase "third culture" originally came from C. P. Snow. In a 1963 essay, Snow suggested that a new caste of thinkers might be emerging to bridge the gulf that so troubled him, a group we might best recognize today as social scientists. See "The Two Cultures: A Second Look," reprinted in C. P. Snow, *The Two Cultures* (Cambridge: Cambridge University Press, 1993). In his own use of the term "third culture," however, Brockman admitted he was improvising, borrowing

only loosely from Snow. In the 1990s, Brockman wrote, "literary intellectuals are not communicating with scientists." Rather, "scientists are communicating directly with the general public." Brockman, *The Third Culture* (New York: Simon and Schuster, 1995).

But to borrow an infamous 1990s expression, it depends on what the meaning of "general public" is. Science book buyers clearly numbered more than enough to make the enterprise profitable. The same is true of right-wing book buyers—just ask the conservative Regnery Publishing. Saying science books sell well, however, is far different from saying they reached a truly mass public, or even that their readership in any way constituted a representative sample of America's nearly 300 million citizens. (Conversation with Matthew Nisbet, August 2008.)

Still, Brockman was right that the 1990s represented a kind of golden age for popular book writing on science, works that forged a culture of enthusiastic, science-loving readers. We should know, because we're part of that culture. We grew up reading those works—especially those by Gould and Dawkins—and drew inspiration from them in much the same way scientists (and science-ists) from the generation preceding ours drew inspiration from Sagan. A decade or more later, we still talk knowingly about Dawkins's "selfish genes" and "memes," and Gould's "non-overlapping magisteria." We know the shared language and use it almost unthinkingly—and to quote C. P. Snow, "that is what a culture means."

It's not easy, then, for us to knock the "third culture." Peering back now, though, it's possible to see negatives as well, at least among individual writers.

43   *"The third culture consists of"*: Brockman, *The Third Culture*.

44   *"virus of the mind"*: Richard Dawkins, "Viruses of the Mind," in Bo Dalhbom, ed., *Dennett and His Critics: Demystifying Mind* (Cambridge, MA: Blackwell, 1993).

44   *"universal acid"*: Daniel C. Dennett, *Darwin's Dangerous Idea* (New York: Simon and Schuster, 1995).

44   *"brights"*: See Richard Dawkins, "The Future Looks Bright," *Guardian*, June 21, 2003, http://www.guardian.co.uk/books/2003/jun/21/society .richarddawkins. See also Daniel C. Dennett, "The Bright Stuff" (op-ed), *New York Times*, July 12, 2003.

44   *public opinion surveys*: National Science Foundation, Science and Engineering Indicators 2008, Chap. 7, "Science and Technology: Public

Attitudes and Understanding," http://www.nsf.gov/statistics/seind08/pdf/c07.pdf.

45   *colleagues across the quadrangle:* Reading *The Third Culture* today, you can't miss a kind of animus against the "other side" much like that which C. P. Snow described among literary intellectuals decades earlier. To quote Nobel laureate physicist Murray Gell-Mann from that book: "Unfortunately, there are people in the arts and humanities—conceivably, even in some of the social sciences—who are proud of knowing very little about science and technology, or about mathematics. The opposite phenomenon is very rare. You may occasionally find a scientist who is ignorant of Shakespeare, but you will never find a scientist who is *proud* of being ignorant of Shakespeare." *The Third Culture*, p. 22.

No wonder science journalist Jonah Lehrer, author of the culture-crossing 2007 book *Proust Was a Neuroscientist*—which relates how a number of nineteenth- and twentieth-century artists and writers appear to have anticipated some discoveries of modern neuroscience—didn't want to be associated with the "third culture" and noted its "serious limitations." Many of its luminaries, he wrote, are "extremely antagonistic toward everything that isn't scientific." Lehrer, *Proust Was a Neuroscientist* (Boston: Houghton Mifflin, 2007), pp. 191–192.

The late Stephen Jay Gould, perhaps the most tolerant of the third culturists toward the separate "magisteria" of religion and the humanities (as he called them), also detested the tone of scientific superiority. "In our increasingly complex and confusing world," wrote Gould in a posthumously published book, "we need all the help we can get from each distinct domain of our emotional and intellectual being . . . *Quilting* a diverse collection of separate patches into a beautiful and integrated coat of many colors, a garment called wisdom . . . sure beats *defeating* or *engulfing* as a metaphor for appropriate interaction." Gould, *The Hedgehog, the Fox, and the Magister's Pox: Mending the Gap Between Science and the Humanities* (New York: Harmony Books, 2003), p. 19.

45   *"postmodernism":* One of the biggest problems with the Science Wars was the apparent reduction of all of the humanities to this vague category. The sociological, historical, philosophical, and cultural study of science is a very worthwhile endeavor. If scholars engaged in such research sometimes take a stance of agnosticism toward the truth claims of science—or consider historic writings in social context, with an emphasis on

revealing implications of language—perhaps that's simply their way of remaining detached from the subject they're studying. But it doesn't necessarily follow that these scholars are "anti-science" or absolute relativists, to the extent of thinking that bedrock scientific concepts like gravity are a mere matter of opinion. It's doubtful that anyone beyond a few academic dilettantes and provocateurs ever really believed this. Stanley Aronowitz, founding editor of *Social Text* and an oft-attacked personage in the Science Wars, has himself written that "the critical theory of science does not refute the results of scientific discoveries since, say, the Copernican revolution or since Galileo's development of the telescope." Aronowitz, "The Politics of the Science Wars," in Andrew Ross, ed., *Science Wars* (Durham: Duke University Press, 1996).

When it comes to the field of science and technology studies (STS), in fact, much scholarly work in the area lends itself not to left-wing attacks on science but rather to defenses of science from forms of abuse prevalent on the political right. To cite just one example, leading science-studies scholar Sheila Jasanoff's 1990 book, *The Fifth Branch: Science Advisers as Policymakers* (Cambridge: Harvard University Press, 1990), presents a potent critique of demands for unreasonable levels of scientific certainty before political decisions can be made, especially when it comes to protecting public health and the environment.

45 *first major fusillade in the so-called Science Wars:* Paul R. Gross and Norman Levitt, *Higher Superstition: The Academic Left and Its Quarrels with Science* (Baltimore: Johns Hopkins University Press, 1994).

45 *"not enemies but friends":* Ibid., p. 2.

46 *"dominant mode of thinking about science":* Ibid., p. 4.

46 *"intellectual dereliction":* Ibid., p. 239.

46 *"hiring, firing, and promotion":* Ibid., p. 255.

46 *ivory-tower war chant:* The famed adage from Columbia University political scientist Wallace Sayre is irresistible here: "Academic politics is the most vicious and bitter form of politics, because the stakes are so low."

46 *Alan Sokal:* For the original paper submitted to *Social Text*, and Sokal's explanation and further musings, see Alan Sokal, *Beyond the Hoax: Science, Philosophy, and Culture* (Oxford: Oxford University Press, 2008).

47 *"physical 'reality'":* Sokal, *Beyond the Hoax*, p. 9.

47 *a special issue: Social Text* response to Alan Sokal, *Lingua Franca*, July–August 1996, http://www.physics.nyu.edu/faculty/sokal/Social Text_reply_LF.pdf.

47 *front page of the* New York Times: Janny Scott, "Postmodern Gravity Deconstructed, Slyly," *New York Times,* May 18, 1996.

47 *next sally came in 1998:* Edward O. Wilson, *Consilience: The Unity of Knowledge* (New York: Vintage Books, 1998).

47 *"partly fuse":* For a powerful critique of the Wilsonian project, see D. Graham Burnett, "A Dream of Reason" (review of *Consilience*), *American Scholar,* Vol. 67, No. 3 (Summer 1998).

48 *scientific land grab:* Among Wilson's most prominent critics was fellow third culturist and Harvard professor Stephen Jay Gould, who objected that some complex phenomena (and fields) will never "reduce" to simpler explanations. A perfect example is ethics, an arena in which the last thing we want is to be tied to physics via our biology. True, one can survey practices that human societies have tolerated or welcomed, and then link some of those to biological underpinnings, but there will always have to be a non-scientific moral judgment about what's right and wrong.

And Gould launched other critiques: Unexpected "emergent" properties in complex systems could never be predicted from simpler components, and moreover, "contingency" in a historical sense—chance occurrences that have a large effect in, say, the course of evolution—also thwarts the Wilsonian program.

Beyond any intellectual merits, Gould also argued that Wilson was being insensitive or even a tad insulting to other disciplines. He demonstrated "an undiminished belief in the superiority of science, and a devaluing based on misunderstanding the aims and definitions pursued by other forms of knowledge and inquiry—an assumption that cannot forge the kind of allegiances he presumably hopes to establish with scholars in the humanities." See Gould, *The Hedgehog, the Fox, and the Magister's Pox*, p. 217.

48 *ultimately irrelevant:* Not surprisingly, Gould also saw the Science Wars as unproductive, exaggerated, and unhelpful. He added this hardly reassuring gloss: "Tell most scientists about the 'science wars'—and I have tried this experiment at least fifty times—and they will stare back at you in utter disbelief. They have never encountered such a thing, never read anything about it, and don't care to interrupt their work to find out. Oh yes, the occasional savvy scientist who pals around in urban intellectual circles may engage the 'wars' and get pissed off . . . But most of my colleagues know nothing at all about the war being supposedly conducted

in (or against) their name." Gould, *The Hedgehog, the Fox, and the Magister's Pox*, p. 102.

48   *"that zeitgeist is unrecognizable":* Sokal, *Beyond the Hoax*, p. xv.

48   *the real enemy at the gate:* So then why, at a time of attacks on science in Congress, did some scientists instead see the enemy within their own academic ranks? The late New York University sociologist of science Dorothy Nelkin had an interesting reading of the phenomenon. Noting that "normally, scientists are slow to respond to political pressures," she remarked that in contrast, the Science Wars had been swift and characterized by great intensity and vituperation. The attacks on the left-wing critics of science, she suggested, might thus be considered a form of scapegoating, one with its roots in the declining influence of science as the cold war came to a close and public scrutiny and skepticism increased. In this challenging context, some more traditionally minded scientists wanted to return to the days when no one questioned them and everyone loved them, and so strolled across the quadrangle and lashed out at their most proximate critics. Yet not only could they not unstir what the 1960s, 1970s, and 1980s had stirred, but the endeavor was fundamentally misled. Nelkin noted: "There are many threats to scientific rationality these days—from religious fundamentalists, right-wing politicians, nativists, and other antiliberal forces. Attacking fellow academics is, of course, easier, but it is grossly misdirected. It is strategically misguided as well. . . . At a time when academic institutions are generally under siege, dividing the academy into warring factions in this way is extraordinarily counterproductive." Nelkin, "The Science Wars: Responses to a Marriage Failed," in Ross, ed., *Science Wars*.

48   *1996 Telecommunications Act:* Our discussion draws on a report by Common Cause, "The Fallout from the Telecommunications Act of 1996: Unintended Consequences and Lessons Learned," 2005; and on Robert McChesney, *Rich Media, Poor Democracy: Communication Politics in Dubious Times* (Urbana: University of Illinois Press, 1999).

50   *English departments:* William Deresiewicz, "Professing Literature in 2008," *The Nation*, March 11, 2008, http://www.thenation.com/doc/20080324/deresiewicz.

50   *"the attack on science has always been* our *game":* Robert Wilson, "Reason in the Sun," *American Scholar*, Vol. 74, No. 3 (Summer 2005), p. 4.

50   *"intellectuals have stopped being in the vanguard":* Bruno Latour, "Why Has Critique Run out of Steam?" *Critical Inquiry,* Vol. 30, No. 2, http://criticalinquiry.uchicago.edu/issues/v30/30n2.Latour.html.

Chapter 5

53   *a grassroots initiative called ScienceDebate2008:* We described the initiative in S. Kirshenbaum, C. Mooney, S. L. Otto, et al., "Science and the Candidates," *Science,* Vol. 320, No. 5873, p. 182, April 11, 2008. See also Shawn Lawrence Otto and Sheril Kirshenbaum, "Science on the Campaign Trail," *Issues in Science and Technology* (Winter 2009), pp. 27–28.

53   *Science matters:* The official ScienceDebate2008 statement wasn't *quite* so simple. It was the following: "Given the many urgent scientific and technological challenges facing America and the rest of the world, the increasing need for accurate scientific information in political decision making, and the vital role scientific innovation plays in spurring economic growth and competitiveness, we call for a public debate in which the U.S. presidential candidates share their views on the issues of The Environment, Health and Medicine, and Science and Technology Policy."

54   *falling behind in science and innovation:* National Academy of Sciences, *Rising Above the Gathering Storm: Energizing and Employing America for a Brighter Economic Future* (Washington, DC: National Academies Press, 2007).

54   *only six of those exchanges:* League of Conservation Voters press release, January 30, 2008, http://www.lcv.org/newsroom/press-releases/laurie-david-joins-national-groups-in-delivering-petitions-to-the-media.html.

54   *ScienceDebate2008 found its invitation declined:* For more details, see Otto and Kirshenbaum, "Science on the Campaign Trail."

55   *leader who shows a deep appreciation of science:* During the campaign, Obama began ramping up his science-policy capacity at around the same time he answered the fourteen questions posed by Science Debate2008. Those answers—including a pledge to restore science funding, address climate change and spur renewable energy investments, and reverse the Bush administration's executive order limiting federal funding for embryonic stem cell research—drew great applause from the scientific community, and before long an unprecedented number of Nobel laureate endorsements. Obama also released the list of science ad-

visers who had drafted his responses. It was an impressive group whose membership included former Clinton administration National Institutes of Health director and Nobel laureate Harold Varmus, former American Association for the Advancement of Science president Gilbert Ommen, and recent Nobel laureate Peter Agre; they strongly suggested that an Obama administration would take scientific advice very seriously. In an early October 2008 letter to the National Academy of Sciences, Obama further assured the scientific community that he would quickly appoint a presidential science adviser to take with him to Washington.

56 *"[They] were terrified":* Shawn Otto, e-mail communication, January 14, 2009.

56 *a strategy of studied political detachment:* One exception proves the rule: During the 1964 election, a group of scientists dubbed Scientists and Engineers for Johnson-Humphrey helped damage the campaign of Republican candidate Barry Goldwater considerably by denouncing his dangerous tone of nuclear aggression, charges that drew widespread attention at the time. However, the group fell apart afterward. See Daniel S. Greenberg, *Science, Money, and Politics* (Chicago: University of Chicago Press, 2001).

56 *"[scientists] don't run for office":* Daniel S. Greenberg, "Absent from Politics, as Usual: Scientists and Engineers," January 5, 2008, http://chronicle .com/review/brainstorm/greenberg/absent-from-politics-as-usual-scientists -and-engineers.

57 *Scientists look at the world and see order:* See Daniel Yankelovich, "Winning Greater Influence for Science," *Issues in Science and Technology* (Summer 2003), http://www.issues.org/19.4/yankelovich.html. In the essay, Yankelovich outlines the "profound difference in worldviews" that separates scientists from the rest of our society—and by extension, from the politicians it elects. This section draws upon and expands Yankelovich's delineation of the "difference in worldviews."

59 *only 8 percent of elected officials:* Susanne B. Haga, "Congress Needs Scientific Schooling," *Detroit Free Press*, January 28, 2009.

59 *scientists and politicians regularly reenact the problem of the "two cultures":* The conservative Canadian politician Preston Manning has observed that scientists and politicians are as "ships passing or colliding in the night." See Manning, "Communicating Effectively with Politicians," speech delivered June 8, 2007, http://www.manningcentre.ca/en/news_ article/43.

59 *"game theory" research involves sports:* This anecdote is reported in Cornelia Dean, "Physicists in Congress Calculate Their Influence," *New York Times*, June 10, 2008.

60 *"No one in Congress senses the need for science in their daily lives":* The Keystone Center, "Science and Technology Policy in Congress: An Assessment of How Congress Seeks, Processes, and Legislates Complex Science and Technology Issues," April 2008, http://www.keystone.org/spp/documents/Final_report6092_4_2008.pdf.

61 *"framers" of policy issues:* Daniel Yankelovich, "Winning Greater Influence for Science," *Issues in Science and Technology* (Summer 2003), http://www.issues.org/19.4/yankelovich.html.

61 *"source-oriented communicators" and "receiver-oriented communicators":* Manning, "Communicating Effectively with Politicians."

62 *contributed scientific language:* The snippet written by Sagan in Carter's 1981 farewell address was a visionary passage remarking that "the same rocket technology that delivers nuclear warheads has also taken us peacefully into space. From that perspective, we see our Earth as it really is— a small and fragile and beautiful blue globe, the only home we have. We see no barriers of race or religion or country." This episode is related in Keay Davidson, *Carl Sagan: A Life* (New York: John Wiley and Sons, 1999), p. 354.

64 *"I made a last ditch pitch":* Shawn Otto, e-mail communication, January 14, 2009.

65 *Obama campaign broke the ice:* Again, see Otto and Kirshenbaum, "Science on the Campaign Trail."

65 *The field is wide open:* In pursuing these further strategies, scientists may first have to become reconciled to the fact that if they really want to bring about political change, then at least in some cases they will have to start playing the reward-and-punishment game, just like everyone else does. So far, scientists have shied from direct electoral engagement, and for a bevy of understandable reasons. Yet it's clear that by identifying scientists with political talents, training and supporting them, and ultimately electing more of them to Congress, science and its supporters could have a beneficial influence on the entire institution.

At present, for instance, Congress's three physicists—Rush Holt (D–NJ), Vernon Ehlers (R–MI), and Bill Foster (D–IL)—are strangers in a strange land. But if such congressional scientists were more numerous, who can doubt that their expertise would filter into decision making far more widely, slowly changing the anti-science culture of the Congress?

It's not merely that we need more congressional scientists; we need more legislators who take science seriously, regardless of background or party affiliation. Some of the greatest friends of science in Congress have been non-scientists: longtime Republican House Science Committee chair Sherwood Boehlert, for instance, or Democratic representative Henry Waxman, who unearthed many of the Bush administration's most egregious science-related abuses.

To increase the number of people like this, it will first be essential to have a comprehensive data set on the science policy positions of all elected representatives, so that their stands on core scientific issues can be identified and rated. This is an approach already employed by numerous interests groups, ranging from the Christian Coalition to the League of Conservation Voters. The experiment has been tried for science as well: In 1996, a group called Science Watch organized a system to rate members of Congress based on their science-related performance. However, when the scorecard emerged, Democrats generally earned considerably higher ratings than Republicans, leading—all too predictably—to charges of politicizing science. Rather than carrying on unintimidated, the science community retreated; no further scorecards from Science Watch were forthcoming.

The enemies and cultural competitors of science have shown few such compunctions. For example, as has now been well documented, fossil-fuel interests for many years adopted a strategy not unlike that of the tobacco industry, sponsoring a sweeping campaign to sow doubts about the linkage between human-induced greenhouse gas emissions and rising global temperatures. And their cause was abetted by sympathetic legislators like Oklahoma Republican senator James Inhofe, who in 2003 dubbed global warming "the greatest hoax ever perpetrated on the American people." Yet with some exceptions, American scientists have been responding to such assaults with one (or both) hands tied behind their backs.

How can we increase the number of legislators who take science seriously? Expanding the ScienceDebate experiment to a wide range of Senate and House races would help. The more science gets injected into elections, the more we'll know what kinds of representatives we're getting when it comes to science policy.

But more could be done. Through the formation of auxiliary groups, those who care about science could directly take on politicians with the most outrageous anti-science stances, such as Inhofe. They could also organize to elect better candidates—including more scientists—to public office and make sure that representatives know there are consequences for attacking scientists and undermining scientific knowledge.

We're not advocating new or foreign strategies; we're just describing what everyone who wants to be politically influential actually does.

In fact, it's possible to go further. Why not form a nonpartisan science political action committee, or PAC, devoted to funding candidates who are either scientists themselves or who make science a strong priority and have good records on science issues? With adequate funding, the PAC might select, say, five or ten members or candidates to support each election cycle. If there's a desire to be really aggressive (and we have mixed feelings about this strategy), it could also target science "bad guys"— climate change deniers, officials who promote manufactured scientific controversies, anti-evolutionists, and the like—who deserve to be un-elected and give campaign funds to their opponents.

Some might argue that in taking such actions, science would be sacrificing its objectivity. Certainly this would represent a new level of political mobilization, with associated risks. Yet there's no doubt the PAC concept has been proven successful at influencing policy. Consider a group like Ocean Champions—a 501(c)(4) organization with a connected political action committee, dubbed Ocean Champions PAC. It's the first such national organization focused solely on preserving the oceans and oceanic wildlife, and it has been markedly successful so far.

The strategy of the group has been simple: Develop a broad, bipartisan base of supporters to cultivate political champions for ocean conservation in the U.S. Congress and in key states. In other words, the group aims to build the political capital necessary to ensure healthy oceans. This requires policy change, so Ocean Champions provides direct contributions to candidates, mobilizes voters, and lobbies. By studying the most effective special-interest groups, the group identified eight hallmarks of success, and not surprisingly, electoral involvement headlines the list. In this PAC model, pro-ocean members of Congress benefit from funding, votes, and good media coverage back home, while Ocean Champions is afforded a better relationship with the members in offices they support. Everyone—especially the oceans—wins.

PACs are the brass knuckles side of politics: They should only be used to support the greatest science champions or to attack the worst enemies, and should be organized by wealthy individuals rather than by broad scientific institutions, which will rightly want to maintain more distance between themselves and direct electioneering. But there's no avoiding the reality that for scientific information to have its maximal impact, scientists must understand what motivates those in the policy world to act. They must speak the language of politics, know its rules, and adapt to the culture of Congress, including, in some cases, being willing to fight hard when there are no other choices.

There's every reason to believe that science can become much more influential in politics—and if we're going to get the science-related policies we need in the future on the science-related problems that matter most, then it *must*. Moreover, the ScienceDebate initiative has already demonstrated that scientists are able to organize themselves politically, at least behind the prospect of a science policy debate.

Some of the steps we've outlined in this note go further, and not all scientists will want to embrace them. But if we spur the science community as a whole toward greater political outreach, starting with Science Debate and building from there, different actors can set their own comfort levels. Most important is to have the motivation, the sense of momentum—the "yes, we can" conviction that it's time for those who care about science to make a difference, and the recognition that you can't do so without taking any risks.

65   *let's give the last word to Otto:* Shawn Otto, e-mail communication, January 14, 2009.

Chapter 6

67   *"all the way up to the executive editor":* Interview with Rick Weiss, July 21, 2008.

67   *pulled off the beat:* As this book went to press, we learned not only of the (at least temporary) reemergence of the *Post* science page, but also of a new reorganization of the remaining *Post* science-coverage capacity under a single editor. See Cristine Russell, "Washington Post Pools Its Resources," *Columbia Journalism Review* (The Observatory), March 6, 2009, http://www.cjr.org/the_observatory/washington_post_pools_its_reso.php.

68   Boston Globe: See Cristine Russell, "Globe Kills Health/Science Section, Keeps Staff," *Columbia Journalism Review* (The Observatory), March 4, 2009, http://www.cjr.org/the_observatory/globe_kills_healthscience_ sect.php.

68   *as their business model collapses:* For a comprehensive overview of the problem of the newspaper today, see Paul Starr, "Goodbye to the Age of Newspapers (Hello to a New Era of Corruption)," *New Republic*, March 4, 2009.

68   *"The Internet . . . has now surpassed":* Pew Center for People and the Press, "Internet Overtakes Newspapers as News Outlet," December 23, 2008, http://people-press.org/report/479/internet-overtakes-newspa-pers-as-news-source.

68   *Washington, DC, bureaus are vanishing:* John McQuaid, "The Demise of the Washington News Bureau," *American Prospect Online*, September 19, 2008, http://www.prospect.org/cs/articles?article=the_demise_of_ the_washington_news_bureau.

68   *As for books, they're out, too:* And still, that's not all. Some newspapers are going under entirely. And other suffering or vanishing areas of coverage include regional reporting—especially in a government watchdog capac-ity—and the arts, including music criticism and theater reviews. See Starr, "Goodbye to the Age of Newspapers."

68   *shrank by nearly two-thirds:* Cristine Russell, "Covering Controversial Science: Improving Reporting on Science and Public Policy," 2006 Working Paper, Joan Shorenstein Center on the Press, Politics, and Pub-lic Policy, http://www.hks.harvard.edu/presspol/research_publications/ papers/working_papers/2006_4.pdf.

69   *exercise and fitness:* Ibid.

69   *one-third of 1 percent of coverage:* We first noticed these data thanks to Matthew Nisbet's blog, *Framing Science,* March 17, 2008, http://science blogs.com/framing-science/2008/03/if_you_watch_five_hours_of_ cab.php. The study in question is from the Project for Excellence in Journalism, "The State of the News Media 2008," http://www.state ofthenewsmedia.org/2008/narrative_cabletv_contentanalysis.php?cat= 1&media=7.

69   *members of the National Association of Science Writers:* Russell, "Covering Controversial Science."

70   *scientists and journalists had problems connecting:* This chapter covers the broad reasons that scientists and journalists don't necessarily get along.

For a wonderfully helpful guide that advises scientists on how to interact with the press, see Richard Hayes and Daniel Grossman, *A Scientist's Guide to Talking with the Media: Practical Advice from the Union of Concerned Scientists* (New Brunswick, NJ: Rutgers University Press, 2006).

70   *have long operated as "two cultures":* We're hardly the first to use this analogy; see, for example, ibid., p. 1. But we can affirm that the divide between scientists and journalists is a textbook Snowean phenomenon in which two intellectually driven groups share very different core assumptions and practices, which in turn make their interactions tense or even perilous. In fact, the science-media divide is a fairly direct modern descendant of the one Snow fretted about, characterized by scientists thinking like scientists and writers thinking like writers.

70   *primary source of science content:* National Science Foundation, Science and Engineering Indicators 2008, Chap. 7, "Science and Technology: Public Attitudes and Understanding," http://www.nsf.gov/statistics/seind08/pdf/c07.pdf.

70   *set the broader media agenda:* As noted in Michael Weigold, "Communicating Science: A Review of the Literature," *Science Communication*, Vol. 23, No. 2 (December 2001), pp. 164–193.

70   *only about 1 percent of front-page stories:* Project for Excellence in Journalism, "How Different Is Murdoch's New *Wall Street Journal*?" April 23, 2008, http://journalism.org/node/10769.

70   *scientists disapprove:* We should concede that not every survey of scientists' views of the media paints a completely dire picture. Consider a 2005–2006 poll of 1,354 stem cell researchers and epidemiologists in the United States, Japan, Germany, France, and the United Kingdom. The survey found that 46 percent of respondents felt their media encounters had had a "mostly positive" impact on their scientific careers; just 3 percent felt the impact to have been "mostly negative." In addition, 57 percent of respondents were "mostly pleased" by their most recent media experience. The study is Hans Peter Peters et al., "Interactions with the Mass Media," *Science,* Vol. 321, July 11, 2008.

Still, scientists' trepidations about the media persist: Even among these more optimistic scientists, nine in ten still worried about the "risk of incorrect quotation" in journalistic stories and eight in ten fretted about the "unpredictability of journalists." Our impression after interacting with many scientists is similar: When it comes to the media, above all they seem deathly afraid of being misquoted, misrepresented, or baited into conflicts with colleagues. There are reputations at stake, and after a negative encounter with a haphazard journalist, the scientist never forgets.

70   *2004 survey of the members of the Union of Concerned Scientists:* Hayes and Grossman, *A Scientist's Guide to Talking with the Media,* p. 2.

71   *Darwin "was hardly even a scientist":* Malcolm Jones, "Who Was More Important: Lincoln or Darwin?" *Newsweek,* July 7–14, 2008. This stereotype of the pocket-protector-armed scientist, with flaring eyebrows, little sense of how to stay in touch with the rest of the world, and little interest in doing so, seems highly prevalent among political and general-interest journalists in particular. In a 1997 Freedom Forum report, 62 percent of the journalists interviewed for the study agreed that scientists are "so intellectual and immersed in their own jargon that they can't communicate with journalists or the public." Jim Hartz and Rick Chappell, "Worlds Apart: How the Distance Between Science and Journalism Threatens America's Future," First Amendment Center, Freedom Forum, 1997.

71   *"In the newsrooms I know":* Andrew Revkin, "Climate Change as News: Challenges in Communicating Environmental Science," in J. C. DiMento and P. M. Doughman, eds., *Climate Change: What It Means for Us, Our Children, and Our Grandchildren* (Boston: MIT Press, 2007), pp. 139–160.

71   *journalists . . . often pounce on some "hot" result:* See Susan Dentzer, "Communicating Medical News—Pitfalls of Health Care Journalism," *New England Journal of Medicine* (Perspective), Vol. 60, No. 1, January 1, 2009.

71   *the public can experience media "whiplash":* See Andrew Revkin, "Climate Experts Tussle over Details: Public Gets Whiplash," *New York Times,* July 29, 2008.

71   *Should you drink more coffee, or less:* See Nicholas Bakalar, "Coffee as a Health Drink? Studies Find Some Benefits," *New York Times,* August 15, 2006.

71   *media reports have provided contradictory answers:* In fact, it has been seriously suggested that most published scientific research findings are false—a finding that, if *true,* would certainly explain much science journalism whiplash. See John P. A. Ioannidis, "Why Most Published Research Findings Are False," *PLoS Medicine,* Vol. 2, No. 8 (August 2005), pp. 696–701.

71 *angles or frames:* As noted in Weigold, "Communicating Science," pp. 164–193.

72 *the notion of "balance":* For a further deconstruction of media "balance," see Chris Mooney, "Blinded by Science: How 'Balanced' Coverage Lets the Scientific Fringe Hijack Reality," *Columbia Journalism Review* (November–December 2004), http://cjrarchives.org/issues/2004/6/mooney-science.asp.

72 *the few remaining global warming "skeptics":* For media "balance" on the climate issue and its pernicious consequences, see Max Boykoff and Jules Boykoff, "Balance as Bias: Global Warming and the U.S. Prestige Press," *Global Environmental Change,* Vol. 14 (2004), pp. 125–136. Later, Max Boykoff found that the "balance" problem in U.S. media coverage of climate change had significantly improved. See his "Flogging a Dead Norm? Newspaper Coverage of Anthropogenic Climate Change in the United States and United Kingdom from 2003 to 2006," *AREA,* Vol. 39, No. 2 (2007), pp. 1–12.

73 *2001 survey of 744 Dutch scientists:* See Jaap Willems, "Bringing Down the Barriers: Public Communication Should Be Part of Common Scientific Practice," *Nature,* Vol. 422, April 3, 2003, p. 470. For survey results in detail, see http://www.bio.vu.nl/WillemsinNature.pdf.

73 *No self-respecting journalist would agree to this:* The scientists in question seem to expect that journalists will practice the kind of high-level quality control that exists in scientific journals, which generally have very slow turnaround times. But journalism isn't science, and shouldn't be—and any scientist who demands such practices is making an error of category.

73 *a low priority compared with many other issues:* See Pew Research Center, "Economy, Jobs Trump All Other Policy Priorities in 2009," January 22, 2009, http://people-press.org/report/485/economy-top-policy-priority. Global warming ranked twentieth among public priorities, behind not only the economy and health care but such matters as "moral decline" and "lobbyists."

73 *a "he said, she said" controversy during the 1990s:* See Boykoff and Boykoff, "Balance as Bias."

73 *has since fallen again into a decline:* See Max Boykoff data, Oxford Environmental Change Institute, late 2008, http://www.eci.ox.ac.uk/research/climate/mediacoverage.php. The data show a worldwide decline in newspaper attention to climate problems.

75   *the biggest behemoths:* By the time you read this, the constellation of media giants may already have changed. It is, as media critic Mark Crispin Miller puts it, "always growing here and shriveling there, with certain of its members bulking up while others slowly fall apart or get digested whole . . . [but] the overall Leviathan itself keeps getting bigger, louder, brighter, forever taking up more time and space, in every street, in countless homes, in every other head." Mark Crispin Miller, "What's Wrong with This Picture?" *The Nation*, December 20, 2001.

75   *"conglomeration" . . . "consolidation":* The terms "concentration," "consolidation," and "conglomeration" are often tossed around indistinguishably in the media context. However, for our purposes, "consolidation" or "concentration" of ownership refers to developments in a particular sector of the media—for instance, television channel ownership is concentrated if one or several companies own the large majority of nationwide TV stations. "Conglomeration," however, refers to major media companies owning properties in many different sectors—film, television, radio, books, and so on. We rely for the distinction on Robert W. McChesney, *Rich Media, Poor Democracy: Communication Politics in Dubious Times* (Urbana: University of Illinois Press, 1999).

75   *programming to the least common denominator:* See Eric Klinenberg, *Fighting for Air: The Battle to Control America's Media* (New York: Henry Holt, 2007).

76   *fragmentation:* Again, we want to thank Matthew Nisbet for directing our attention to this problem and the research that explores it.

76   *the broadcast networks:* For the transition from broadcast news to cable, and the implications for an informed public, see Markus Prior, "News vs. Entertainment: How Increasing Media Choice Widens Gaps in Political Knowledge and Turnout," *American Journal of Political Science*, Vol. 49, No. 3 (July 2005), pp. 577–592. In a series of lectures delivered with Chris in 2007 and 2008, Matthew Nisbet frequently pointed out, based on Prior's work, that twenty years ago, if you sat down to watch the news in the evening, these were your four choices—ABC, NBC, CBS, PBS.

77   *The consequences are profound:* Again, for an analysis of how media fragmentation decreases political literacy among those members of the public not much interested in the subject, see Prior, "News vs. Entertainment."

77   *employed over a dozen staff science writers:* Russell, "Covering Controversial Science."

78   *without addressing or even raising the broader issue:* For a typical example, see EvolutionBlog, "The Trouble with Science Journalism," January 22, 2009, http://scienceblogs.com/evolutionblog/2009/01/the_trouble_with_science_journ.php.

78   *"I say good riddance":* See "Science Journalism: When Things Get Rough, You Find Out Who Your Real Friends Are," February 23, 2009, http://scienceblogs.com/intersection/2009/02/science_journalism_when_things.php.

Chapter 7

82   *plotlines dependent upon the supernatural and the paranormal:* William Evans, "Science and Reason in Film and Television," *Skeptical Inquirer* (January–February 1996).

82   *planet's core stops spinning:* An occurrence that, if it really did happen, would likely destroy us all and pose a vastly bigger risk than the film's alleged microwave threat from space. No, waves of the sort produced by your microwave could not destroy the Golden Gate Bridge. For critiques of the science of *The Core*, see Phil Plait's movie review, http://www.badastronomy.com/bad/movies/thecore_review.html. See also the discussion by Sidney Perkowitz in *Hollywood Science: Movies, Science, and the End of the World* (New York: Columbia University Press, 2007), pp. 85–86. Finally, see Michael Barnett, Heather Wagner, Anne Gatling, et al., "The Impact of Science Fiction Film on Student Understanding of Science," *Journal of Science Education and Technology*, Vol. 15, No. 2 (April 2006), in which the authors use *The Core* as a case study.

82   *"unofficial curriculum of society":* Interview with Marty Kaplan, August 28, 2008.

82   *one-third of the top fifty biggest film moneymakers:* Perkowitz, *Hollywood Science*, p. 12.

83   *the only audience group that sees past the veneer:* See David Kirby, "Science Consultants, Fictional Films, and Scientific Practice," *Social Studies of Science*, Vol. 33, No. 2 (April 2003), pp. 231–268.

83   *contested on a scientific level:* As discussed in ibid.

83   *call on a scientist to consult:* See Scott Frank, "Reel Reality: Science Consultants in Hollywood," *Science as Culture*, Vol. 12, No. 4 (December 2003).

83   *"hold hands through space suits":* This episode is related in Kirby, "Science Consultants."

83   "not boring": University of Southern California Annenberg School, "Enter the Entertainment Initiative," informational materials.

83   *"imperative to capture and hold attention":* Interview with Marty Kaplan, August 28, 2008.

84   *the San Andreas fault:* See Perkowitz, *Hollywood Science,* p. 85.

84   *NBC's . . . four-hour miniseries* 10.5: For a scientific critique of *10.5,* see Sid Perkins, "What's Wrong with This Picture? Educating via Analyses of Science in Movies and TV," *Science News,* October 16, 2004.

84   *CBS's* Category 7: For a further parsing of nonsense in such movies, see Barnett, Wagner, Gatling, et al., "The Impact of Science Fiction Film on Student Understanding of Science," May 16, 2007 lecture, http://frontrow.bc.edu/program/barnett/.

84   *films like* The Core *and* Volcano: See ibid.

85   *Hollywood's medical plots . . . are legion:* A 2008 analysis by the Kaiser Family Foundation and USC's Norman Lear Center found that from 2004 through 2006, out of the top ten most highly rated scripted (i.e., not "reality") shows on TV, 59 percent of episodes "had at least one health-related storyline," with "storyline" defined as three or more lines of dialogue on a health topic. See Sheila T. Murphy, Heather J. Hether, and Victoria Rideout, "How Healthy Is Prime Time? An Analysis of Health Content in Popular Prime Time Television Programs," Kaiser Family Foundation, September 2008.

85   *"misrepresentation of CPR on television shows":* Susan J. Diem, John D. Lantos, and James A. Tulsky, "Cardiopulmonary Resuscitation on Television: Miracles and Misinformation," *New England Journal of Medicine,* Vol. 334, No. 24, June 13, 1996, pp. 1578–1582.

85   *movies generally "show scientists as idiosyncratic nerds":* Quoted in Perkowitz, *Hollywood Science,* p. 172.

85   *one in six scientists was depicted a villain:* Scientist hero characters also emerge from the products of Hollywood (there are several in *Jurassic Park,* for instance). But we shouldn't confuse counterexamples with a counterargument. In the entire corpus of Hollywood film, it's undeniable there are a whole lot of freaky, geeky, and even evil scientists.

85   *one in ten got killed:* George Gerbner, Larry Gross, Michael Morgan, et al., "Television Entertainment and Viewers' Conceptions of Science," University of Pennsylvania Annenberg School of Communications, July

1985. In another study, surveying 100 films made through the late 1980s and examining previous research on the subject, Lehigh University's Stephen L. Goldman found that science and technology "have been depicted largely negatively in popular films of all genres." See Goldman, "Images of Technology in Popular Films: Discussion and Filmography," *Science, Technology, and Human Values*, Vol. 14, No. 3 (Summer 1989), pp. 275–301.

85 *stereotypical views of scientists held by children:* See, for example, Gayle A. Buck, Diandra Leslie-Pelecky, and Susan K. Kirby, "Bringing Female Scientists into the Elementary Classroom: Confronting the Strength of Elementary Students' Stereotypical Images of Scientists," *Journal of Elementary Science Education*, Vol. 14, No. 2 (Fall 2002), pp. 1–9.

85 *"They might say the person was too 'normal'":* Quoted in Jonathan Knight, "Hollywood or Bust: Last Month, a Handful of Scientists Who Have Toyed with the Idea of Writing for the Movies Were Given a Masterclass by Tinseltown's Finest," *Nature*, Vol. 430, August 12, 2004.

86 *a long literary tradition:* See Jon Turney, *Frankenstein's Footsteps: Science, Genetics, and Popular Culture* (New Haven: Yale University Press, 1998).

86 *knowledge leads the scientist to play God:* Or as Victor Frankenstein puts it in Shelley's novel: "Learn . . . by my example, how dangerous is the acquirement of knowledge, and how much happier that man is who believes his native town to be the world, than he who aspires to become greater than his nature will allow."

86 *such depictions go all the way back to Fritz Lang:* We can also detect the Frankenstein mythology in such late nineteenth- and early twentieth-century novels as H. G. Wells's *The Island of Dr. Moreau* and Aldous Huxley's *Brave New World*, both of which also became films.

86 *mad scientist Rotwang builds an evil robot:* As scientist and film enthusiast Sidney Perkowitz summarizes the plot: "Boy meets girl, boy loses girl, boy builds girl." Perkowitz, *Hollywood Science*, p. 7.

86 *The paradigmatic modern example of the evil scientist trope . . . E.T.:* There are many other such films, often linked to the biomedical sciences and especially to the subject of cloning. One thinks of films like 2005's *The Island*—in which the doctor running the clone complex has a "God complex"—but the same trope appears in flicks ranging from *Jurassic Park* to *Star Wars* (especially episodes 2 and 3) to 1995's *Batman Forever*, in which mad scientist Edward Nigma ("The Riddler") develops a device to extract victims' thoughts and intelligence, making him smarter but

ultimately contributing to his mental breakdown. As *Slate* magazine put it after surveying nearly a century of "mad scientist" films: "What do cinematic images of scientists say about cultural attitudes toward scientific progress? They are usually about science as a source of anxiety, scientists as outsiders and oddballs, research as very likely to get into the wrong hands, and scientific institutions as dangerous places to be. Never mind that cinema depends on technological progress—this is one of the great unresolved contradictions of popular culture." Christopher Frayling, "Spawn of Frankenstein: Mad Scientists in the Movies," *Slate*, May 9, 2006, http://www.slate.com/id/2140772/.

Possibly that contradiction arises in part as the legacy of the late 1960s and 1970s, a period that instilled doubt about science and its societal benefits and costs among many intellectuals, including those in the entertainment industry. Possibly it goes back still further, descending from the battles between C. P. Snow and the "literary intellectuals" who had such a negative view of technology-dependent industrialization. *Frankenstein* was a product of the British Romantic movement.

These are undoubtedly important antecedents and influences; but with film and television we must always bear in mind that the bottom line involves not ideology, but profits—or as Stephen Colbert has so felicitously put it, getting "asses in seats." So if these depictions of scientists recur, it's likely they serve some purpose, even if it's one as narrow as predictability—giving audiences more of what they already know and have thus come to expect.

87 *"being rational is considered the opposite of being creative":* Interview with Matthew Chapman, August 20, 2008.

87 *"threateningly intelligent":* Interview with Joe Petricca, September 5, 2008.

87 *maximum degree of ichthyological realism possible:* See Alison Abbot, "The Fabulous Fish Guy," *Nature,* Vol. 427, February 19, 2004, pp. 672–673.

87 *"word that comes to mind is* serendipitous": Interview with Martin Gunderson, August 27, 2008.

87 *hectoring annoyance:* Not every scientist is overly literal-minded or unable to grasp the exigencies of storytelling. But the generalization isn't entirely without merit. For instance, Joe Petricca describes an experiment in which the American Film Institute set up a workshop to train a group of scientists in the art of screenwriting. "They would have these incredible, fascinating, you-can't-believe-they're-true scientific ideas,

stuff you can't even make up," he remembers, "but not a character in sight, not a person, no nothing." But "the granularity of science, the specificity of science, doesn't help the story," Petricca said. "The story is about honesty, the story is about emotion—even if it's a ridiculous story, or a comedy story. Those are the things that will speak to an audience." Interview with Joe Petricca, September 5, 2008.

88   *"The natural world is fascinating":* Richard Dawkins quoted in Andrew Pollack, "Scientists Seek a New Movie Role: Hero, Not Villain," *New York Times*, December 1, 1998.

88   *movie producer Michael Crichton:* Despite his late-life attacks on climate science, there can be no doubt that Michael Crichton was a great innovator and had massive influence upon the depiction of science in Hollywood and in popular culture. As Chris has argued, we shouldn't let one late-life misjudgment totally cloud our image of him. See Mooney, "The Crichton Effect: A Chief Designer of the Image of Science in America Passes," *Science Progress,* November 11, 2008, http://www.scienceprogress.org/2008/11/the-crichton-effect/.

88   *four important rules of movies:* Michael Crichton, "Ritual Abuse, Hot Air, and Missed Opportunities," 1999 AAAS annual meeting lecture, Anaheim, California, http://www.crichton-official.com/speech-scienceviews media.html.

89   *seek out constructive consulting roles:* There's at least some evidence suggesting Hollywood science-consultant numbers are actually on the rise—or at least, they were as of 2003, according to a study published in that year by University of Manchester science-communication scholar David Kirby. Kirby, "Science Consultants."

89   *reported that the experience was a very enjoyable one:* Frank, "Reel Reality."

89   *and that they made a real difference:* There are risks in science consulting, such as getting used as a rubber stamp whom filmmakers can cite to show that they did indeed run things by an expert. The scientifically ludicrous *Volcano,* for instance, had a consultant, and *The Core* director Jon Amiel has boasted, "We wanted to actually put some science in the science fiction." Quoted in Cindy White, "Director Jon Amiel Spearheads a New Journey to the Center of the Earth," *Sci-Fi Weekly,* March 24, 2003, http://www.scifi.com/sfw/interviews/sfw9571.html. But it's also clear that those scientists who work or consult with Hollywood can learn a great deal about how to be an effective go-between, connecting two worlds and two cultures.

89   *attempts to cast scientific leaders . . . in the role of out-and-out villains:*
Frank, "Reel Reality."

89   *invited onboard by those at the head of film projects:* Ibid.

89   *By the time a science consultant arrives:* Ibid.

89   *aware of the realities and constraints of filmmaking:* As an example, consider Marty Kaplan's Norman Lear Center, whose "Hollywood Health
and Society" project—funded by the Centers for Disease Control and the
National Institutes of Health—focuses on medical content in television
dramas. One central role played by the project is to take calls from television writers who have medical questions and quickly put them in contact with on-call experts: Over 200 such inquiries were answered in the
year 2006. See Lauren Movius, Michael Cody, Grace Huang, et al., "Motivating Television Viewers to Become Organ Donors," *Cases in Public
Health Communication and Marketing* (June 2007). "Without trying to
wag a finger, without saying that there is a compromise between profit—
which is what the entertainment industry is about—and storytelling,
there are often intriguing solutions that can be both accurate and live
within the parameters of the storytelling situation you're in," explains
Kaplan. "We don't succeed because people want to be good guys . . . We
succeed because writers and producers and network executives have come
to the point of view that if you can be accurate, without a cost to the entertainment value and story structure and so on, it's probably a good
thing to do." (Interview with Marty Kaplan, August 28, 2008.)

Sensitivity of this type is critical, because Hollywood is already massively overlobbied by scores of interest groups that monitor the depictions of all sorts of subjects, and all have their own grievances and wants.
The "Hollywood Health and Society" project teaches another lesson,
too, a more practical one. It is this: Although those unfamiliar with the
industry tend to think first about ways of influencing blockbuster films,
the truth is that it's possible to have a far greater and more immediate
impact through television.

The reason is sheer numbers. Studios see tens of thousands of film-
scripts and script ideas per year, and even after winnowing these down
dramatically and putting some small percentage of projects into development, only one in twenty ideas that survive the first cut gets produced. By the time the call has been made to bring a particular film
project all the way to production, "there have been 7 million meetings,
script conferences, notes," says Kaplan. "And in at the end walks a sci-

entist or consultant, and says, 'No, it can't work like that'? The notion of throwing out key elements of the story is not very welcome." (Interview with Marty Kaplan, August 28, 2008.)

Television is different. The audience is still very large: During the week of October 22, 2007, the ten leading network prime-time shows drew in 15 to 21 million viewers apiece. Murphy, Hether, and Rideout, "How Healthy Is Prime Time?" But these shows—like *House* and *Grey's Anatomy*—have to produce weekly episodes, often retaining for that purpose regular consultants who work intensively with the writers week in and week out to offer ideas and feedback.

Consider the successful CBS series *Numb3rs*, which features a non-nerdy mathematician (played by David Krumholtz) who helps his FBI agent brother solve crimes. Stanford University mathematician Keith Devlin, who consulted for the show during its first season, was aware that creators Ridley and Tony Scott very much wanted to have a scientist-mathematician hero character. He explains his motivation for helping out thusly: "I just knew, if it was successful as a series, it would have a huge impact on the public perception of mathematics." That certainly did occur—during the 2005–2006 season *Numb3rs* garnered an average of 13 million viewers per night. (Movius, Cody, Huang, et al., "Motivating Television Viewers to Become Organ Donors.") Devlin adds: "Once audiences get used to the fact that you can have good science intelligently portrayed, that raises the bar and the expectations." (Interview with Keith Devlin, September 5, 2008.)

USC's Norman Lear Center worked with *Numb3rs* writer J. David Harden to provide information for a January 27, 2006, episode, entitled "Harvest," which involved the subject of organ donation. (Movius, Cody, Huang, et al., "Motivating Television Viewers to Become Organ Donors.") The episode, which was viewed by 13.36 million people, ended with a kind of teachable moment in which the characters discuss how important it is to become an organ donor; afterward, in surveys posted on the show's Web site and fan sites, viewers expressed themselves as more likely to sign up to be a donor after seeing the episode. As Harden has commented: "I'm not naive—we all understand TV has a big impact. Just consider the size of the audience for our show: Eleven million people and upwards watching Friday nights. You definitely live with a sense that there's some responsibility incumbent upon you in the face of that audience."

90   *a Hollywood film had a massive impact:* This isn't the only example of Hollywood depictions influencing politics. In 1998, two blockbuster films—
Armageddon and *Deep Impact*—dramatized the risk to the planet from a
collision with a so-called near-Earth object (an asteroid in one film, a comet
in the other). Although one could quibble with some of the specific details
in these films—the asteroid in *Armageddon* is unrealistically large, for instance—there's no doubt that both treated a scenario that doesn't merely lie
within the realm of theoretical possibility but has occurred repeatedly in the
past history of the planet. These treatments resulted in dramatically increased public awareness and closely coincided with (and may have helped
trigger) the launch of NASA's Near-Earth Object Program, which now
tracks such risks. For further details, see Kirby, "Science Consultants."

90   *growth of student interest in the field:* See Scott Smallwood, "As Seen on
TV: 'CSI' and 'The X-Files' Help Build Forensics Programs," *Chronicle
of Higher Education*, July 19, 2002.

90   *Science and Entertainment Exchange:* See http://www.scienceand
entertainmentexchange.org/index.html.

90   *a positive role model . . . who overcomes numerous obstacles:* Still, none of
this could prevent the need to construct, for the film version of *Contact*,
a love interest for Arroway that doesn't exist in the novel—her semi-
mechanical fling with the preacher Palmer Joss (Matthew McConnaughey).
Some scientists were displeased at the compulsory affair, but, well, that's
Hollywood. See Pollack, "Scientists Seek a New Movie Role."

91   *earned $171 million worldwide: Contact* earnings figures from Box Office
Mojo, http://www.boxofficemojo.com/movies/?id=contact.htm.

91   *$589 million: Men in Black* earnings figures from Box Office Mojo,
http://www.boxofficemojo.com/movies/?id=meninblack.htm.

91   *$817 million: Independence Day* earnings figures from Box Office Mojo,
http://www.boxofficemojo.com/movies/?id=independenceday.htm.

91   *$178 million: Dante's Peak* earnings figures from Box Office Mojo,
http://www.boxofficemojo.com/movies/?id=dantespeak.htm.

91   *$122 million: Volcano* earnings figures from Box Office Mojo,
http://www.boxofficemojo.com/movies/?id=volcano.htm.

92   *"genetic determinism":* See David Kirby, "The New Eugenics in Cinema:
Genetic Determinism and Gene Therapy in *Gattaca*," *Science Fiction
Studies*, Vol. 2, Pt. 2 (2000), pp. 193–215.

92   Gattaca *challenges such questionable presumptions:* Alas, *Gattaca* wasn't
successful enough at the box office to count as a model that Hollywood

would want to follow again—it only grossed $12 million domestically. *Gattaca* earnings figures from Box Office Mojo, http://www.boxofficemojo .com/movies/?id=gattaca.htm.

92 *$542 million: The Day After Tomorrow* earnings figures from Box Office Mojo, http://www.boxofficemojo.com/movies/?id=dayaftertomorrow.htm.

92 *those who had seen it were significantly more worried:* Anthony Leiserowitz, "Before and After *The Day After Tomorrow*," *Environment* 46 (2004), pp. 22–37. Leiserowitz also found that the film's U.S. viewership within a few weeks of its release was roughly 21 million, or 10 percent of the population. Yet this was not enough to significantly move total public opinion on global warming. That's a sobering consideration, especially in light of yet another of Leiserowitz's findings: The film generated over ten times as much media attention as the 2001 release of the U.N. Intergovernmental Panel on Climate Change's "Third Assessment Report," the definitive scientific study of climate change and its impact, which is released at roughly five-year intervals. In the total media arena, then, *The Day After Tomorrow* made a much bigger splash than the release of a groundbreaking scientific report, but a much smaller one than a sustained politico-media scandal story, such as the Abu Ghraib prison saga.

93 *fifth-highest-grossing political documentary:* Data from Box Office Mojo, http://www.boxofficemojo.com/genres/chart/?id=politicaldoc.htm.

Chapter 8

95 *holding it "hostage":* http://www.wftv.com/news/16798008/detail.html.

95 *"frackin cracker":* http://scienceblogs.com/pharyngula/2008/07/its_ a_goddamned_cracker.php.

96 *"I pierced it . . . with a rusty nail":* http://scienceblogs.com/pharyngula/ 2008/07/the_great_desecration.php.

96 *over 2 million page views per month:* http://www.sitemeter.com/?a= stats&s=sm1pharyngula&r=33.

97 *hardly a monolithic group:* For instance, Hitchens's writings suggest he would disapprove of the desecration of religious symbols. See *God Is Not Great* (New York: Twelve, 2007), p. 11: "I leave it to the faithful to burn each other's churches and mosques and synagogues, which they can always be relied upon to do. When I go to the mosque, I take off my shoes. When I go to the synagogue, I cover my head."

97 *Harris . . . rejects the atheist label:* See Sam Harris, "The Problem with Atheism," September 28, 2007, lecture, http://newsweek.washingtonpost .com/onfaith/sam_harris/2007/10/the_problem_with_atheism.html.

97 *Sam Harris questions . . . tolerating religious moderates:* Sam Harris, *The End of Faith* (New York: Norton, 2004 [paperback]). See, for example, p. 15: "I hope to show that the very ideal of religious tolerance—born of the notion that every human being should be free to believe whatever he wants about God—is one of the principal forces driving us toward the abyss."

97 *"Neville Chamberlain school of evolutionists":* Richard Dawkins, *The God Delusion* (Boston: Houghton Mifflin, 2006); see pp. 66–69. Such charges are then taken up by radicalized followers, as shown in comments like the following from one atheist blog, which introduces the charming term "theistard": "Why should we encourage the malignant tumour of religion? Why, appeasers? Because you people are a bunch of spineless pushovers? Because you appeasers are a bunch of theistard-enablers who seem to be acting more and more like theistards as the days go by?" Quotation from http://www.evolvedrational.com/2008/04/ spineless-appeasers-or-closet-theistard.html.

98 *alarming percentage of our citizens (46 percent):* National Science Foundation, Science and Engineering Indicators 2008, Chap. 7, "Science and Technology: Public Attitudes and Understanding," http://www.nsf.gov/ statistics/seind08/pdf/c07.pdf.

98 *so with the Big Bang:* Ibid.

98 *A 2007 study:* Elaine Howard Ecklund and Christopher P. Scheitle, "Religion Among Academic Scientists: Distinctions, Disciplines, and Demographics," *Social Problems*, Vol. 54, No. 2, pp. 289–307.

99 *wholly dismantled by scientific experts:* See, for instance, Philip Kitcher, *Abusing Science: The Case Against Creationism* (Cambridge: MIT Press, 1982).

99 *aren't really operating on that level:* What the evolution wars really need, in our opinion, is better strategy, better communication, more serious engagement, and the rethinking of assumptions. Frankly, they also need you to write a check to the leading defender of evolution in the country, the Oakland, California–based National Center for Science Education, which does not attack religion and whose director, Eugenie Scott, commented to us as follows on the science-religion question: "It just isn't a matter of either you're an atheist who believes in evolution, or you're a

Christian who believes in special creation. There's all kinds of intermediate positions here that people need to consider, and if your goal is to keep evolution in the schools, help people understand what science is, and why it's such a good way of learning about the natural world, that dichotomization is just starting off in a ten foot deep hole. Why handicap yourself?"

100   *"The appeal of creationism is emotional":* Kenneth Miller, *Finding Darwin's God: A Scientist's Search for Common Ground Between God and Evolution* (New York: HarperCollins, 2002 [paperback]), p. 173.

100   *the essential organizing principle of their lives:* As Miller summarized the viewpoint in ibid., pp. 186–187: "If evolution leads logically to the exclusion of God from a meaningless universe, then evolution must be fought at every opportunity." Added Michigan State University philosopher of science Robert Pennock: "Creationists believe that moral value itself is at stake." Pennock, *Tower of Babel: The Evidence Against the New Creationism* (Cambridge: MIT Press, 1999), p. 311.

The internal papers of the intelligent design movement further demonstrate that the animus against evolution isn't really driven by science— it's far bigger than that. The notorious "Wedge Document," from the intelligent design–promoting Discovery Institute, accuses Charles Darwin of joining Karl Marx and Sigmund Freud in a triumvirate of thinkers who "portrayed humans not as moral and spiritual beings, but as animals or machines who inhabited a universe ruled by purely impersonal forces and whose behavior and very thoughts were dictated by the unbending forces of biology, chemistry, and environment." The consequences of this materialism, the document alleges, have been "devastating"—and that's why the Discovery Institute seeks "nothing less than the overthrow of materialism and its cultural legacies." For the text of the "Wedge Document," see http://ncseweb.org/creationism/general/wedge-document.

100   *quality science and supernatural beliefs are irreconcilable:* The idea that science and faith are pretty much doomed to conflict is one of the New Atheists' central doctrines. As *End of Faith* author Sam Harris has put it: "The conflict between religion and science is inherent and (very nearly) zero-sum. The success of science often comes at the expense of religious dogma; the maintenance of religious dogma always comes at the expense of science." Harris, "Science Must Destroy Religion," *Huffington Post,* January 2, 2006, http://www.huffingtonpost.com/sam-harris/science-must-destroy-reli_b_13153.html.

100   *"damaging to the well-being of the human race":* This scene at City College was reported in Cornelia Dean, "Scientists Speak Up on Mix of God and Science," *New York Times,* August 23, 2005.

100   *historical scholarship:* For a historically sophisticated analysis of the relationship between science and religion that eschews ideology and debunks both the strong "conflict" and strong "harmony" narratives, see John Hedley Brooke, *Science and Religion: Some Historical Perspectives* (New York: Cambridge University Press, 1991).

100   *contradicts Hauptman's simplistic assertion:* And for that matter, any assertion that the relationship is, or even could be, simple or black and white. The intricacy of the science-religion question arises from the dramatic ways in which both have changed over the centuries, to say nothing of the diversity of world faiths, the diversity *within* faiths, and the interaction of both science and religion with the political and economic realities prevailing in different nations at different times. For all of these reasons, the truth is, as science historian John Hedley Brooke has written, "There is no such thing as *the* relationship between science and religion. It is what different individuals and communities have made of it in a plethora of different contexts." Ibid., p. 321. For another historical account that does not support either a "conflict" or a "harmony" thesis, see David C. Lindberg and Ronald L. Numbers, eds., *God and Nature: Historical Essays on the Encounter Between Christianity and Science* (Berkeley: University of California Press, 1986).

100   *leading lights of the scientific revolution and the Enlightenment:* These scientists had no problem throwing off some incorrect notions inherited at least in part from religious traditions (perhaps most notably an Earth-centered cosmology) while retaining an underlying faith. As the eighteenth-century English chemist Joseph Priestley put it, scientific advancement could serve as the "means under God of extirpating all error and prejudice, and of putting an end to all undue and usurped authority in the business of religion as well as of science." Quoted in Brooke, *Science and Religion,* p. 25.

101   *"conflict" narrative:* For Draper and White see David C. Lindberg and Ronald L. Numbers, Introduction to *God and Nature: Historical Essays on the Encounter Between Christianity and Science* (Berkeley: University of California Press, 1986).

101   *"Thou didst":* Biblical quotations from *The New Oxford Annotated Bible with the Apocrypha,* ed. Herbert G. May and Bruce M. Metzger (New York: Oxford University Press, 1973).

101 *diseases have come to be understood as naturally caused:* By 1800, Brooke notes, "It had become less acceptable to ascribe illness to divine warning or punishment." *Science and Religion*, p. 155.

102 *thus Benjamin Franklin invented the lightning rod:* This point concerning lightning is made in Robert T. Pennock's testimony at the Dover evolution trial, http://www.talkorigins.org/faqs/dover/day3am.html.

102 *proof of a designer's active hand:* Such was the case made in Anglican priest William Paley's influential 1802 book, *Natural Theology*, which advanced the famous "argument from design." In yet another indication of the complexity of the science-religion relationship, natural theology drove a considerable amount of scientific inquiry, inspiring parsons and priests to become students of nature and seek out new evidence of what they perceived to be God's work. See Brooke, *Science and Religion*, pp. 192–225.

102 *religion would have to retreat:* And this time the retreat was particularly painful, because although Darwin at first made little mention of the implications of his theory for "man and his origins," everyone could tell where this line of thought carried. We weren't so special; and nature, featuring constant bloody competition for survival and a long history of extinctions, didn't sound much like the kind of system that a benevolent, loving God would design.

102 *Darwinian revolution:* Rather than forever unseating religion, then, Darwin is perhaps better understood as having shown the power of a scientific methodology that did not admit of any role for the miraculous or supernatural causation when it came to explaining the workings of things. As University of Wisconsin science historian Ronald Numbers puts it: "By the mid 19th century, students of nature, scientists, were almost unanimous—even among some of the creationists—in their conviction that appeals to the supernatural had no place in doing science. It didn't mean you couldn't believe in the supernatural and many of them did, but if you appealed to a miracle, then that was cheating." Interview with Ronald Numbers, October 20, 2008.

102 *major Anglican clergymen:* Brooke, *Science and Religion*, pp. 41, 293–294.

102 *"The more we know of the fixed laws of nature":* Quoted in ibid., p. 271. Today, the idea of trying to preserve a role for religion in the shrinking space of natural explanation has come to be called the "God of the gaps" approach, and disdained by serious religious thinkers. It's simply a losing

strategy; to quote Kenneth Miller, it turns God into a mere "magician" and a bumbling one at that, constantly meddling around with nature and designing organisms fated to later go extinct. This is why many theologians and religious adherents have such a problem with "intelligent design": They think it trivializes God at least as much as it impugns science.

And yet none of what we know from science excludes a different religious possibility: That some creator set in motion the laws of nature, and now—albeit in a scientifically undetectable way—acts through them. "The discovery that naturalistic explanations can account for the workings of living things neither confirms nor denies the idea that a Creator is responsible for them," writes Miller (*Finding Darwin's God,* p. 268). John Haught, a Catholic theologian at Georgetown University, even argues that had a camera been present at the scene of Christ's rising from the dead, it would have recorded nothing. "We trivialize the whole meaning of the Resurrection when we start asking, Is it scientifically verifiable?" says Haught. See Steve Paulson, "The Atheist Delusion," Salon.com, December 18, 2007.

103   *National Academy of Sciences:* See National Academies, *Teaching About Evolution and the Nature of Science* (Washington, DC: National Academies Press, 1998), p. 58, noting: "Religions and science answer different questions about the world. Whether there is a purpose to the universe or a purpose for human existence are not questions for science."

103   *American Association for the Advancement of Science:* As Francisco Ayala, former priest and past president of AAAS, has written, "Science and religion concern nonoverlapping realms of knowledge. It is only when assertions are made beyond their legitimate boundaries that evolutionary theory and religious belief appear to be antithetical." Quoted in Cornelia Dean, "Roving Defender of Evolution, and of Room for God," *New York Times,* April 29, 2008.

Ayala's words invoke an influential postulate about the science-religion relationship that comes from the work of Stephen Jay Gould, who described science and religion as "nonoverlapping magisteria" (NOMA) that cannot conflict due to their separate spheres—a distinction designed to promote "the principled resolution of supposed 'conflict' or 'warfare' between science and religion." Gould, "Nonoverlapping Magisteria," in *Leonardo's Mountain of Clams and the Diet of Worms* (New York: Three Rivers Press, 1998).

As Gould described NOMA: "No conflict should exist because each subject has a legitimate magisterium, or domain of teaching authority—and these magisteria do not overlap . . . The net of science covers the empirical realm: what the universe is made of (fact) and why does it work this way (theory). The net of religion extends over questions of moral meaning and value. These two magisteria do not overlap, nor do they encompass all inquiry (consider, for starters, the magisterium of art and the meaning of beauty). To cite the usual cliches, we get the ages of rocks, and religion retains the rock of ages; we study how the heavens go, and they determine how to go to heaven."

Let us acknowledge that this stance is not unproblematic. For instance, religion cannot be the only source of "moral meaning and value" or else atheists couldn't have any, which is clearly not the case. And many religious believers—especially creationists—would have a lot of trouble restraining themselves from making claims about the empirical realm.

Still, Gould's words express an honest attempt to ensure respect and avoid conflict, and one with "important practical consequences in a world of such diverse passions," for it demands "mutual humility." And note that it is no refutation of the NOMA principle to find cases in which religious believers have transgressed against it by making claims about the natural world. This is something creationists do all the time and in fact, they do by definition. Gould knew people are constantly violating NOMA, but his prescription is that this is where the line ought to be drawn in order to create conditions conducive to overall peace and understanding.

103   not *the same thing as philosophical naturalism:* To elaborate: A philosophical naturalist, in Robert Pennock's definition, "would be someone who says the world as it is in its ultimate reality, its metaphysical reality, is nothing but material natural processes, and there is no supernatural, there is no god, there is nothing beyond." However, this claim is about the ultimate nature of things, and so is not based on science itself. "Science is not in the business of making philosophical metaphysical claims." See the testimony of Robert Pennock in the Dover, Pennsylvania, evolution trial of 2005, http://www.talkorigins.org/faqs/dover/day3am.html. Elsewhere Pennock has written, "Scientists need to recognize and respect, as most do, the limits of methodological naturalism. If individual scientists wish to dive into deeper metaphysical waters, then they should be clear when they are doing so . . . and not suggest that

their conclusions are drawn strictly from within science." Pennock, "God of the Gaps: The Argument from Ignorance and the Limits of Methodological Naturalism," in Andrew J. Petto and Laurie R. Godfrey, eds., *Scientists Confront Creationism: Intelligent Design and Beyond* (New York: Norton, 2008).

104 *"one is setting aside questions about whether the gods"*: Again, see Pennock's testimony in the Dover, Pennsylvania, evolution trial of 2005, http://www.talkorigins.org/faqs/dover/day3am.html. For a more thorough exposition of his views on this question, see his book *Tower of Babel: The Evidence Against the New Creationism* (Cambridge: MIT Press, 1999).

104 *"Science is godless in the same way that plumbing is godless"*: Ibid., p. 282.

104 *collapse the distinction between methodological and philosophical naturalism:* Richard Dawkins is one example of this, but the same goes for another New Atheist, Victor Stenger, whose book title says it all: *God: The Failed Hypothesis. How Science Shows That God Does Not Exist.*

104 *"unequivocally a scientific question"*: Dawkins, *The God Delusion*, p. 59. In fact, Dawkins repeatedly claims that his critiques of the existence of God are "scientific" in nature, rather than philosophical or metaphysical. Or as he puts it at one point in the book, the existence of God is "a scientific question; one day we may know the answer, and meanwhile we can say something pretty strong about the probability" (p. 48). At yet another point in the book, he argues that "the existence of God is a scientific hypothesis like any other" (p. 50).

We're confounded by such claims. If God is a supernatural being, and supernatural agents are, by definition, "not constrained by natural laws" (Pennock, *Tower of Babel*, p. 289), then surely we cannot use science's "methodological naturalism" to know anything about them. That includes testing whether they exist or establishing the probability of such existence.

To further underscore this point, let's examine Dawkins's attempts to refute theologians who claim that God is "simple." To the contrary, Dawkins argues, "however little we know about God, the one thing we can be sure of is that he would have to be very very complex and presumably irreducibly so!" (p. 125). To Dawkins, such complexity makes God highly improbable, and this is one of his central arguments. The reasoning here draws upon Dawkins's previous, important work on how natural selection creates complex organisms, described in books such as his *Climbing Mount Improbable*—but that's precisely the problem. Why should we assume naturalistic arguments about "complexity" are applicable to a supernatural being? And even more important, how could we ever know reliably whether they apply? Certainly we cannot know as much through science.

At another point in *The God Delusion*, Dawkins attempts a similar maneuver, arguing that "a universe with a creative superintendent would be a very different kind of universe from one without. Why is that not a scientific matter?" (p. 55). Simple: because it is untestable by science. We have a hard time seeing how one can examine two universes and control for the existence of a creative superintendent to see how they differ, when the creative superintendent is (again) supernatural and therefore not constrained by natural law.

The point is that while Dawkins's position on God's existence may be influenced by his scientific training and the kind of person it has made him—the thoughts it has engendered—the position itself is not the result of science. To quote Duke University evolutionary scientist Matt Cartmill: "Many scientists are atheists or agnostics who want to believe that the natural world they study is all there is, and being only human, they try to persuade themselves that science gives them grounds for that belief. It's an honorable belief, but it isn't a research finding." Cartmill, "Oppressed by Evolution," *Discover*, Vol. 19, No. 3 (March 1998), pp. 78–83.

None of this is to say that Dawkins is wrong; merely that with such arguments, he's going beyond the realm of science, and shouldn't claim otherwise. As a philosopher, he may well be right. Chris finds Dawkins's "cosmic teapot" argument, cribbed from Bertrand Russell, particularly persuasive. This is the idea that there is no more reason to believe in any supernatural entity than there is to believe in a teapot floating somewhere in the middle of space. We see no reason to believe in something without evidence of its existence—but then, we also recognize that the term "evidence" itself presumes naturalism.

At minimum, then, Dawkins's assertion that the refutation of God's existence can proceed scientifically is highly questionable.

104   *an intellectual error at best:* It's an error of no small consequence for the standing of science in America. Part of the weight of the compatibilist stance with respect to science and religion derives not merely from its philosophical or historical underpinnings (although these are very

strong), but its practical significance. That significance manifests itself in multiple arenas—the legal one, the educational one, and in the broader public sphere, where it serves to maintain the good name of the scientific community and to show that it's open-minded and tolerant.

Let's take the law first. The pro-evolution legal strategy has long rested on the principle that evolutionary science is neutral with respect to religion, unable either to prove or disprove God's existence. This in turn sets up the argument that creationism of course *is* religion and thus illegal to teach in public school science classes under the First Amendment. Precisely such a strategy prevailed in the famous 2005 Dover, Pennsylvania, trial over the teaching of "intelligent design." Far from being a scientific alternative to evolution, ID didn't count as science at all, ruled the Bush-appointed district court judge John E. Jones III. It was a souped-up form of creationism, perhaps, but still ultimately reducible to religion. Therefore, its teaching violated the separation of church and state. As Jones put it in his opinion (relying heavily upon Robert Pennock's testimony): "In deliberately omitting theological or 'ultimate' explanations for the existence or characteristics of the natural world, science does not consider issues of 'meaning' or 'purpose' in the world. While supernatural explanations may be important and have merit, they are not part of science. This self-imposed convention of science, which limits inquiry to testable, natural explanations about the natural world, is referred to by philosophers as 'methodological naturalism' and is sometimes known as the scientific method. Methodological naturalism is a 'ground rule' of science today which requires scientists to seek explanations in the world around us based upon what we can observe, test, replicate, and verify." See http://www.talkorigins.org/faqs/dover/kitzmiller_v_dover_decision.html/.

Thus has the compatibilist position on the relationship between science and religion devastated the anti-evolutionists in court.

And if the compatibilist stance has been crucial in the legal arena, it is equally important in science education, where mini-battles over science and religion are erupting constantly. As Patricia H. Kelley, a geologist at the University of North Carolina–Wilmington, puts it: "In my own teaching experience, which has been mostly in the southern United States, students entering my class frequently have had the preconceived notion that science and religion are incompatible, and that they must either make a conscious decision to reject science or to reject religion."

Kelley, "Stephen Jay Gould's Winnowing Fork: Science, Religion, and Creationism," in Warren D. Allmon, Patricia H. Kelley, and Robert M. Moss, eds., *Stephen Jay Gould: Reflections on His View of Life* (Oxford: Oxford University Press, 2009).

Any teacher dealing with such students on the subject of evolution would be mad to take the New Atheist line with them. Yet teaching evolution from a "methodological naturalist" perspective lets everyone meet on common ground without feeling a threat to their beliefs, because they are not at stake, not on the table. See William W. Cobern, "The Competition of Secularism and Religion in Science Education," Special Supplement, *Religion in the News* (Summer–Fall 2007).

In fact, education researchers have found that defusing the tension over science and religion facilitates learning about evolution. "I submit that anti-religious rhetoric is counter-productive. It actually hampers science education," writes Shawn K. Stover, a biologist at Davis and Elkins College in West Virginia. In Stover's view, students who feel that evolution is a threat to their beliefs will not "*want* to learn," and only reconciliatory discussion can open them up to evolution. Stover, "The Great Divide: How to Resolve the War Between Science and Religion," *eSkeptic*, September 24, 2008.

104   *a nasty bullying tactic:* Insofar as the new atheism strives to reach beyond science's limitations—boundaries that end at the natural world—and claims that it's "scientific" to be an atheist, then it also seeks to turn science into an anti-religious doctrine. In a very religious country like the United States, this would vastly strengthen the claims of anti-science religious conservatives, who strategically blur the distinction between science and atheism in order to lump them together. In a 2007 *New York Times* op-ed, for instance, Senator Sam Brownback (R–KS) wrote that "if evolution means assenting to an exclusively materialistic, deterministic vision of the world that holds no place for a guiding intelligence then I reject it." But evolution *doesn't* mean that: It can't; it simply describes how human beings and other animals came to exist in their current form. Whether God was in some way also involved, perhaps by creating the universe and the laws that ultimately led to our existence through evolution, is a matter that's simply impossible to address on a scientific level.

105   *"faith and science can and should coexist":* Eric Alterman, "Why We're Liberals: The Polls Speak," April 30, 2008, http://www.americanprogress .org/issues/2008/03/alterman_book.html.

105 *accept the teaching of evolution:* We came across this list of religious organizations and their statements on evolution and creationism in Kelley,
"Stephen Jay Gould's Winnowing Fork."

105 *Clergy Letter Project:* See http://www.butler.edu/clergyproject/Christian_
Clergy/ChrClergyLtr.htm.

105 *"the weakness of the religious mind":* Dawkins, *The God Delusion,* p. 16.
The God Delusion amounts to a "coming out" book for atheists, so no
wonder it has riled them so (p. 4). Although its author says he will not
"go out of my way to offend" in the book, neither will he "don kid
gloves to handle religion any more gently than I would handle anything
else" (p. 27). Dawkins's wit is devastating, and his arguments powerful as
well, but that's precisely the point: Why then does he need to express
them with such condescension?

We want to emphasize that New Atheists enjoy freedom of speech.
No one is asking them to be quiet. However, we have every right to
point out the consequences of the divisiveness they are fueling over science and religion.

What's more, we have every right to ask this question: Why is it necessary that the intellectual case for atheism be made without moderation, conciliation, or humility? If atheism has compelling arguments
behind it (as we believe it does), then atheists ought to be the first to
reach out to religious believers on their own terms, seeking to create the
types of dialogue that might convince some of them to reconsider what
they've long held as true.

105 *Sagan subjected all the standard arguments:* Carl Sagan, *The Varieties of
Scientific Experience: A Personal View of the Search for God,* ed. Ann
Druyan (New York: Penguin Press, 2006).

106 *treated the subject of religion respectfully:* As Sagan put it in ibid., "I
would suggest that science is, at least in part, informed worship. My
deeply held belief is that if a god of anything like the traditional sort exists, then our curiosity and intelligence are provided by such a god. We
would be unappreciative of those gifts if we suppressed our passion to
explore the universe and ourselves. On the other hand, if such a traditional god does not exist, then our curiosity and our intelligence are the
essential tools for managing our survival in an extremely dangerous
time. In either case the enterprise of knowledge is consistent surely with
science; it should be with religion, and it is essential for the welfare of
the human species" (p. 31).

106 *imported into the neuroscience arena:* Amanda Gefter, "Creationists De-clare War over the Brain," *New Scientist*, October 22, 2008, http://www.newscientist.com/article/mg20026793.000-creationists-declare-war-over-the-brain.html.

Chapter 9

109 *Watts Up With That defeated Pharyngula:* See http://2008.weblogawards.org/polls/best-science-blog/.

110 *new form of citizen-journalist media—blogging:* A blog (short for "weblog") is a continuously updated Web page, on which entries are posted sequentially with the most recent item at the top of the page. Readers can often participate in an ongoing online conversation by leaving comments.

110 *exploded in popularity and readership:* Several studies have come up with very different estimates of the total number of blogs in existence, in part because it's hard to determine how many of the supposed 100 million-plus blogs out there are truly active. See Technorati, "State of the Blogo-sphere 2008," http://www.technorati.com/blogging/state-of-the-blogo sphere/. But one thing is certain: Blogging is an expanding worldwide phenomenon that has enormous impact on policy, entertainment, and other spheres, and especially the news media.

110 *93 percent of the leading 100 newspapers:* The Bivings Group, "The Use of the Internet by America's Newspapers," December 18, 2008, http://www.bivings.com/thelab/presentations/2008study.pdf.

110 *largest blogs, such as the Huffington Post:* According to the blog-indexing site Technorati, as of late 2006, twenty-two of the top 100 most-linked news and information sources were blogs. Technorati, "State of the Blogo-sphere 2007," http://www.sifry.com/alerts/archives/000493.html.

110 *1,000 science blogs in existence:* Laura Bonetta, "Scientists Enter the Blogo-sphere," *Cell,* Vol. 129, May 4, 2007, pp. 443–445.

110 *where we hang our hats:* Chris has been blogging, with varying degrees of intensity, since the year 2001—shortly after the phenomenon really kicked into gear in the wake of the 9/11 attacks—and has been blogging about science since 2003, when he launched The Intersection. Sheril joined that blog in 2007 while still working within academia and has experienced firsthand the mixed feelings about blogging that currently exist in the ivory tower. The Intersection spent several years on the ScienceBlogs network before moving, in 2009, to Discover Blogs.

111 *leaving other, older sources in the dust:* The Internet's "margin over other sources is large and growing," observes the National Science Foundation. Science and Engineering Indicators 2008, Chap. 7, "Science and Technology: Public Attitudes and Understanding," http://www.nsf.gov/statistics/seind08/pdf/c07.pdf.

111 *back-scratching communities:* For an elaboration of this problem, see Cass Sunstein, *Republic.com* (Princeton: Princeton University Press, 2001).

111 *bridge the traditional "two cultures" divide:* John S. Wilkins, "The Roles, Reasons and Restrictions of Science Blogs," *Trends in Ecology and Evolution,* Vol. 23, No. 8 (August 2008), pp. 411–413.

111 *anti-science forces . . . establish their own Internet hubs:* As Sunstein and colleagues have written, the grouping together of like-minded people (in this case, strong science supporters) can lead to "increased extremism, decreased internal diversity, and greater divisions across ideological lines. These effects should be expected to occur when groups self-sort in purely geographical terms; they should also occur when the sorting occurs in terms of what people read and watch." See David Schkade, Cass Sunstein, and Reid Hastie, "What Happened on Deliberation Day," Working Paper, July 2006, AEI-Brookings Joint Center for Regulatory Studies, http://aei-brookings.org/admin/authorpdfs/redirect-safely.php?fname=../pdffiles/phpb7.pdf.

The famed philosopher Jürgen Habermas has similarly remarked that "in the context of liberal regimes the rise of millions of fragmented chatrooms across the world [tends] to lead to the fragmentation of large, but politically focused mass audiences into a huge number of isolated issue publics." Habermas, "Political Communication in Media Society— Does Democracy Still Enjoy an Epistemic Dimension? The Impact of Normative Theory on Empirical Research" (see his footnote 14), http://www.icahdq.org/Speech_by_Habermas.pdf.

112 *the soundly refuted claim:* See Thomas C. Peterson, William M. Connolley, and John Fleck, "The Myth of the 1970s Global Cooling Scientific Consensus," *Bulletin of the American Meteorological Society* (September 2008), pp. 1325–1337.

112 *their ever-growing influence on the traditional press:* It has become a regular blogospheric occurrence that once a number of sites start buzzing about a topic, the result is increased pressure upon traditional journalists to take up the cry as well. To relate just one example from our own experience, in May 2008 Sheril composed a short blog post entitled

"What's Not Making News," which addressed the topic of ocean acidification and simply read: "Ocean acidification is intimately connected to our changing climate and as important as global warming. We're just not hearing about it in the news enough because the media has all but ignored the problem. So we must make the case that more scientists ought to be exploring the threat, educating the public as to why it matters, and implementing effective policy to mitigate the impact of excess CO_2 in our oceans (and everywhere else)," http://scienceblogs.com/intersection/2008/05/whats_not_in_the_news.php.

The entry was swiftly picked up by major news media sources in print and even inspired a story about ocean acidification that aired during prime time on *ABC News,* featuring the famed marine explorer Sylvia Earle. In a matter of days, a single blog post drew national media attention to a critical environmental issue.

112   *online version of a scientific conference:* Unlike the average conference, though, this dialogue is open to interested non-expert members of the public who can also comment, question the scientists, and participate in the growth of knowledge—all without having to pay upward of $1,000 for airfare, admission, and a hotel stay, which is the typical minimum expenditure required to attend a scientific meeting these days.

112   *comment threads:* As mentioned above, blogs generally allow readers to post comments. A "thread" is simply a long list of these, keyed to a particular blog entry, in which the most recent comment sits lowest on the page.

112   *hundreds of high-level, astute contributions:* Such examples help defuse the skepticism, which exists both within and outside the scientific community, about the value of participating in the blogosphere. We don't much sympathize with those traditionalist scientists who frown on blogging, worrying that unlike professional academic publications, blogs offer no formal peer-review process even as politics, opinion, and attitude abound. There are reports from some young scientists that they've been told to "stop blogging and concentrate on 'real' work" (Wilkins, "The Roles, Reasons and Restrictions of Science Blogs"), yet in fact, all kinds of innovative scientific thinking is being facilitated by the Web, and at least among the best science blogs, a de facto peer-review process also exists thanks to reader commentary and participation.

Thus, although the response to science blogging from the institutional scientific community has been mixed, it appears to be gaining acceptance as a useful tool within the ivory tower. With increasing frequency, prominent scientists can be found sending press releases about their published work to well-established blogs in the hope of achieving wider distribution.

And no wonder: Whereas most scientific journals are expensive to subscribe to and typically difficult to access, blogs are free and openly available to everyone online. Scientific journals have even begun to treat bloggers like traditional science journalists, sending them notifications and embargoed articles before publication, hoping that they will receive attention and commentary in the blogosphere.

112 *user-friendly, and open-access dialogue about science:* For many outside of academia, blogging affords the opportunity to ask experts questions and interact with the science community. In this way, science blogging has at least some potential to break down the "us" versus "them" mentality separating scientists and non-scientists, and to help interested citizens separate fact from hype on critical health and environmental issues such as the alleged link between vaccines and autism.

113 *report back online:* Whenever our friend Duke researcher Vanessa Woods travels to Congo with her husband, Dr. Brian Hare, to study bonobos (an endangered close relative of humans), her blog "Bonobo Handshake" (http://bonobohandshake.blogspot.com) describes her work and encourages readers to participate in conservation. Across the globe we find something similar: Meteorologist Tasmin Gray blogs at "Frozen Cheese" (http://www.frozen-cheese.blogspot.com/) about climate change and the ozone hole from the Halley research station in the Antarctic. And when Sheril traveled to South Africa with conservation biologist Stuart Pimm, she brought Intersection readers along by posting entries about biodiversity and research during the journey.

At the same time, blogging has made the science community, or at least some of its individual members, seem far less imposing and intimidating. For example, University of Southern California physicist Clifford Johnson may write about space exploration one day and about his gardening hobby the next on his blog "Asymptotia" (http://asymptotia .com/). Readers who follow scientist bloggers are constantly reminded that scientists are people, too.

113 *formation of a "blogger coalition":* http://www.sciencedebate2008.com/ www/index.php?id=9.

113 *stumbling over them by accident:* As *Houston Chronicle* science reporter (and blogger) Eric Berger has put it: "The Web is a great medium for scientists to share and discuss results through pre-prints, listservs, wikis and the like. But it's not good at all for engaging a wide segment of the public-at-large in science. The general public, the average member of which is not inclined toward science, is unlikely to stumble upon or patronize Seed or ScienceBlogs. They're busy going to ESPN or Gawker or a host of other Web sites. The great thing about the Internet is that people have absolute choice over the content they consume. The terrible thing about the Internet is that people have absolute choice over the content they consume." Eric Berger, comment to The Intersection, October 29, 2008, http://scienceblogs.com/intersection/2008/10/the_science_writers_lament.php#comment-1180076.

113 *"feeding the beast":* Because of these demands on time and energy, science blogging attracts a particular type of personality. Bloggers must participate every day in an ongoing online dialogue, constantly generate new content, and (perhaps) moderate and answer comments, all while attending to whatever professional and personal responsibilities they may have. Those able to shoulder all this will not necessarily stand as representative of their disciplines: Indeed, according to Technorati, 57 percent of U.S. bloggers are male and 42 percent are age eighteen to thirty-four; 75 percent have college degrees; 42 percent have been to graduate school; and more than half have an annual household income of over $75,000. Technorati, "State of the Blogosphere 2008," http://www.technorati.com/blogging/state-of-the-blogosphere/who-are-the-bloggers/. These bloggers are not exactly a snapshot of the general public, or even the science community at large. For instance, in our experience men vastly outnumber women in the science blogosphere.

113 *favors polemicism over nuance:* A related drawback lies in the fact that much science blogging takes a pervasively critical, debunking attitude. Perhaps the dominant mode of discourse is to take some misconception or bit of misinformation, skewer it, and explain the truth of things. Such debunking performs an important service, but at the same time, it leads to a hypercritical orientation, in which science bloggers are always slicing and dicing bad logic and bad facts (and even bad grammar). Once again, such fare won't necessarily have much appeal to any audience other than an already science-centered one.

Meanwhile, the number of visitors to blogs who take an active role in their comments section is generally a small fraction of those who read the site silently (the so-called lurkers), and those who do post comments often generate nastiness and attacks rather than constructive dialogue.

Consider the once-popular ScienceBlog called Next Generation Energy (http://scienceblogs.com/energy/), on which Sheril wrote a post about rising to the energy challenges of the twenty-first century. The thread of commentary that followed was hijacked by an unrelenting climate change denier known as "LudHunter," who responded to every other participant with skepticism and personal attacks against what he termed "gaia-worshipping masturbation in the echo chamber," http://scienceblogs.com/energy/2008/07/a_reaction_commensurate_to_the.php. Eventually one of "Ludhunter's" profane responses was held for moderation because of its content, and as a result he composed a post on his personal blog called "Spiked by the tyrant snatch on scienceblogs.com," calling moderation a "classic Goebbels move," http://ludhunter.wordpress.com/2008/07/. He subsequently apologized, explaining he had "downed 3 Jack and cokes just before, and was gettin' ornery," but the dialogue (long since abandoned by Sheril) continued for over a week. In the end, "LudHunter" got the last word by telling the final objector to "get off the doom bandwagon before she soils her polyester shirt with tofu and bong resin." Not surprisingly, after this exchange, Next Generation Energy never regained the same traffic numbers.

114   *2 million unique visits:* http://sitemeter.com/?a=stats&s=sm1pharygula&r=33.

114   *likely also the most alienating one:* Even within the science-blogging community, many have been offended by, and polarized over, unending fights over science and religion. Rob Knop, a physicist and former ScienceBlogs contributor who happens also to be a Christian, found himself pilloried on the site after responding to the claim that science and religion are always mutually exclusive. In 2007, he ended a long post by explaining: "Anything that happens as real 'miracle,' as real divine intervention, happens through human agency, so there's no need to invoke the divine if you want to explain the mechanics of it, which is after all what science is all about. But the *meaning* of it—well, that's not even a valid question often in science, but that doesn't mean it's not a valid question to *people*, and that's why we have humanities and theology and other forms of intellectual endeavors that aren't science," http://scienceblogs.com/interactions/2007/03/so_why_am_i_a_christian_specif_1.php.

Here Knop sought to bridge the culture gap by reconciling two different ways of thinking and communicating about the world—but even though his post generated nearly 200 comments, and some readers were willing to ask questions, many atheists and religious fundamentalists alike called him delusional, indoctrinated, and worse.

114 *half a million visitors per month:* http://www.sitemeter.com/?a=stats&s=s36wattsup.

Chapter 10

117 *sounded the alarm:* National Academy of Sciences, *Rising Above the Gathering Storm: Energizing and Employing America for a Brighter Economic Future,* Committee on Science, Engineering, and Public Policy (Washington, DC: National Academies Press, 2007). But note that an early version of the report came out in late 2005. See National Academies press release, "Broad Federal Effort Urgently Needed to Create New, High-Quality Jobs for All Americans in the 21st Century," October 12, 2005, http://www8.nationalacademies.org/onpinews/newsitem.aspx?RecordID=11463.

118 *large and very welcome increases:* For details on the science-funding provisions of the economic stimulus bill of early 2009, see American Association for the Advancement of Science, "Final Stimulus Bill Provides $21.5 Billion for Federal R&D," http://www.aaas.org/spp/rd/stim09c.htm.

118 *more Ph.D.s . . . each year:* National Science Foundation, "2007 Records Fifth Consecutive Annual Increase in U.S. Doctoral Awards," November 2008, http://www.nsf.gov/statistics/infbrief/nsf09307/.

118 *more than any other nation in the biomedical research arena:* Organization for Economic Co-operation and Development, Main Science and Technology Indicators, biannual series, 2008.

118 *total government-funded research and development:* Ibid.

118 *employ the most scientists:* Ibid.

118 *chief source of valuable new patents:* National Science Foundation, Science and Engineering Indicators 2008, Chap. 6, http://www.nsf.gov/statistics/seind08/pdf/c06.pdf. See Figure 6–38, based on Organisation for Economic Co-operation and Development patent data.

118 *publish vastly more peer-reviewed research:* National Science Foundation, Science and Engineering Indicators 2008, Chap. 5, http://www.nsf.gov/statistics/seind08/pdf/c05.pdf.

119   *"Neither . . . scientific literacy":* National Academy of Sciences, *America's Lab Report: Investigations in High School Science* (Washington, DC: National Academies Press, 2005).

119   *Stein's own words:* Ben Stein, "All Kinds of 'Wonder'-Ful," *Entertainment Weekly,* May 24, 1991, http://www.ew.com/ew/article/0,,314414,00 .html.

120   *Miley Cyrus probably seems a lot more relevant to their lives:* Our public policies don't help when it comes to making science relevant to students. The No Child Left Behind Act, for instance, requires states to test students' factual knowledge of science rather than encouraging a more comprehensive appreciation for the subject. Ursula Goodenough, a professor of biology at Washington University in St. Louis who participated in a 2005 state science-standards-review process conducted by the Fordham Institute, has made the case that this will lead to an overemphasis on "teaching the test": "As things now stand, K–12 students go into science classes and hear about cells one day and atoms another day, but lack any opportunity or guidance for integrating these understandings into larger contexts . . . most students find science classes tedious and boring and drop out as soon as they've met the requirements." Tony Fitzpatrick, "Tell a Story: Teaching Science Should Have a Narrative Component, Goodenough Says," *Washington University St. Louis Record,* March 24, 2006.

120   *a 2007 study by the Urban Institute:* B. Lindsay Lowell and Harold Salzman, "Into the Eye of the Storm: Assessing the Evidence on Science and Engineering Education, Quality, and Workforce Demand," Urban Institute, 2007, http://www.urban.org/UploadedPDF/411562_Salzman_ Science.pdf.

120   *Who are these lost scientists?:* And possibly the students who lose that sense of wonderment with science that they may once have had, as a result of being continually forced to memorize from college textbooks that seem to increase in girth every few years—so much that "elephantiasis of the textbook" is now a subject actually fretted over in the published scientific literature. See R. C. Kerber, "Elephantiasis of the Textbook," *Journal of Chemical Education,* Vol. 65 (1988), pp. 719–720.

120   *median time spent getting a Ph.D.:* University of Chicago, National Opinion Research Center, Survey of Earned Doctorates, "2006 Doctorate Recipients from United States Universities: Summary Report," http://www.norc.org/projects/survey+of+earned+doctorates.htm.

120 *median age at the time of doctorate receipt:* Ibid.

121 *pressing on can become a heavy financial burden:* In a 2003 article for the *Scientist,* Daniel S. Greenberg vividly illustrated the problem: "Consider the economic fates of two bright college graduates, Jane and Jill, both 22. Jane excels at a top law school, and after graduation three years later, is wooed and hired by a top law firm at the going rate—$125,000 a year, with a year end bonus of $25,000 to $50,000.

"Jill heads down the long trail to a PhD in physics, and after six Spartan years on graduate stipends rising to $20,000 a year, finally gets her degree. Tenure-track jobs appropriate to her rigorous training are scarce, but, more fortunate than her other classmates, she lands a good postdoc appointment—at $35,000 a year, without health insurance or professional independence. Three years later, when attorney Jane is raking in $150,000 a year, plus bonuses, Jill is nail-biting over another postdoc appointment, with an unusually ample postdoc recompense of $45,000 per annum. Medicine and business management similarly trump science in earning power." Greenberg, "The Mythical Scientist Shortage," *Scientist,* Vol. 17, No. 6, March 24, 2003, p. 68.

121 *record number of science and engineering Ph.D.s:* National Science Foundation, Directorate for Social, Behavioral, and Economic Sciences Info Brief, November 2008, http://www.nsf.gov/statistics/infbrief/nsf09307/nsf09307.pdf.

121 *five straight years of increases:* The 2007 figures are the most recent available data at the time this book is going to press. Here it may help to anticipate some objections. It is often asked, "But aren't all these new Ph.D.s from other countries than the U.S.?" The truth is that although the rate of increase is bigger for non-U.S. citizen degree recipients, all doctorates are on the rise. See ibid. The entire body of Ph.D.s is growing; moreover, it's important to bear in mind that many foreign students remain in the United States to work after obtaining their doctorates.

Whether all of these scientists are "enough" to keep us competitive is exceedingly difficult to say. However, it's definitely incorrect to suggest (as so many regularly do) that our numbers of researchers are in decline at the moment.

121 *48,000 . . . "postdocs":* National Postdoctoral Association, "Postdoctoral Scholars Fact Sheet," http://www.nationalpostdoc.org/atf/cf/%7B89152 E81-F2CB-430C-B151-49D071AEB33E%7D/PostdocScholarsFact sheet.pdf.

121   *an average of 1.9 years:* Ibid.

121   *Fifty-eight percent:* Ibid.

121   *34 percent:* Ibid.

121   *from 74 to 44 percent:* Ibid. The decline was from 60 to 31 percent at re-
      search universities.

122   *only 7 percent:* B. L. Benderly, "The Incredible Shrinking Tenure Track,"
      Science Careers, July 2, 2004, http://sciencecareers.sciencemag.org/career_
      development/previous_issues/articles/3150/the_incredible_shrinking_
      tenure_track.

122   *"I'm a recent PhD graduate":* See http://www.scienceprogress.org/2008/
      12/where-are-the-grad-students/#comment-3944.

123   *Every year:* Data collected before the U.S. recession was declared in
      2008.

123   *more than three times as many . . . graduates:* Lowell and Salzman, "Into
      the Eye of the Storm."

123   *not if it only provides technical scientific expertise:* As the Urban Institute
      study put it: "In our interviews with engineering managers . . . rarely, if
      ever, do they say they are unable to find graduates with the requisite
      technical skills but rather the 'shortage' is of engineers with communica-
      tion, management, interpersonal and other soft skills."

123   *"soft skills":* Interview with Bill Bates, December 10, 2008.

123   *situations of ridiculous competition:* Matters have become particularly
      strained in the biomedical arena, where the constriction of opportunity
      for the youngest scientists has become a cause of outrage. As funding
      levels for the National Institutes of Health have declined sharply since
      2003, young scientists have been disproportionately punished by a sys-
      tem where having a prior history of receiving research funds makes all
      the difference—where, in short, the rich get richer, and the young get
      sacrificed. According to *Science* magazine, the average age for receiving a
      first grant from the NIH is forty-two. See Jocelyn Kaiser, "The Graying
      of NIH Research," *Science,* Vol. 322, No. 5903 (November 7, 2008), pp.
      848–849. Less than three decades ago, 25 percent of the main indepen-
      dent research grants in the biomedical field went to scientists under the
      age of thirty-five; today, it's less than 3 percent.
        The NIH comprises an enormous slice of the financial pie for science—
      disbursing nearly $30 billion annually, about six times as much funding
      as the National Science Foundation—so its particular funding history
      explains many of the troubling trends we're seeing today. As Congress

doubled the NIH budget between 1998 and 2003, many new Ph.D. positions became available as principal investigators could afford to support multiple students. In turn, established faculty members had increased opportunities to publish and submit better grant proposals. By the time funding leveled off, they had a demonstrated history of winning awards based on the boom. See "A Broken Pipeline? Flat Funding of the NIH Puts a Generation of Science at Risk," a Follow-Up Statement by a Group of Concerned Universities and Research Institutions, March 2008, http://www.brokenpipeline.org/brokenpipeline.pdf.

The consequence today is that significantly more older scientists are finding funding compared with their younger colleagues. Again and again, grant money is going to those with a proven track record rather than those pushing novel, potentially groundbreaking ideas. Perhaps the situation was best summed up by Bush administration National Institutes of Health director Elias A. Zerhouni, who has worried that we must not "eat our seed corn." See Zerhouni, "NIH in the Post-Doubling Era: Realities and Strategies," *Science,* Vol. 314, No. 5802 (November 17, 2006), pp. 1088–1090.

124 *falling behind in science:* None of what we say in this chapter should be taken as an attempt to discount this grave concern. Consider: Although the United States still leads the world in total science and engineering Ph.D. production—and though our total number of Ph.D.s produced is also increasing, at least for the moment—China's rate of increase is far greater as it approaches us from behind, a fact suggesting we may be ceding our lead and that it (and other nations) are catching up. Only about one-third of bachelor's degrees attained in the United States are in science and engineering; by contrast, more than half of Chinese first degrees are in these fields. See National Science Foundation, "Science and Engineering Indicators 2008," Chap. 2, http://www.nsf.gov/statistics/seind08/pdf/c02.pdf. If China continues to expand its science and engineering programs at its current pace, it may overtake U.S. Ph.D. production very soon; and in fact, by other measures its technological competitiveness is even closer to (or outstrips) ours. See Science|Business, "China Now Leading the U.S. in Technological Competitiveness," February 7, 2008, http://bulletin.sciencebusiness.net/ebulletins/showissue.php3?page=/548/2732/9918.

124 *training a cadre of communication and outreach experts:* An additional intriguing idea would be to pursue precisely the kind of synthesis that this

book represents. Why not team up science graduates who wish to pursue outreach with recently minted science reporters who've just graduated from specialized journalism programs? Even though science-reporting jobs are vanishing from the traditional media, we find journalism schools producing more specialized science reporters than before. At present, science writer Cristine Russell notes, "the journalism pipeline for new science reporters is bigger than ever before," with thirty graduate-level programs across the country specializing in training them. Russell, "Covering Controversial Science: Improving Reporting on Science and Public Policy," 2006 Working Paper, Joan Shorenstein Center on the Press, Politics, and Public Policy, http://www.hks.harvard.edu/presspol/research_publications/papers/working_papers/2006_4.pdf. It's yet another pipeline mismatch that ought to be turned to our advantage, a "two cultures" collaboration just waiting to happen.

125   *rethinking* scientist *education:* This is not to say that this way of thinking shouldn't arc back to the opening stretches of the science pipeline as well—it certainly should. For instance, K–12 education should foster a more realistic portrayal of scientists. Schools should regularly partner with scientific institutions that can provide visiting scientists who are fun and colorful (not boring!) and willing to come talk to classes regularly. They should ensure that science teachers are prepared for a stronger curriculum and enthusiastic to keep our youngest students engaged.

Given that most six-year-olds already love subjects such as dinosaurs and space exploration, science shouldn't be a difficult sell. So by all means, as the National Academies suggested in the *Gathering Storm* report, let's recruit 10,000 teachers to educate 10 million minds. If the United States can succeed at keeping students engaged, it will be an investment producing exponential results. A firm understanding of science will foster the next generation of scientific leadership and literacy.

By the time students reach high school, we should provide rigorous challenges and also make sure students do not lose sight of science's intimate relationship with other subjects. And it's imperative that we shatter the false caricature that all scientists are out-of-touch nerds who can't get a date. Schools should continue to bring in professionals from many walks of life who are engaging, interesting, and involved in changing the world. Hip, fun, trailblazing research pioneers are everywhere; we ought to start celebrating them. Let's make names like Bonnie Bassler and Pardis Sabeti (Google them) as recognizable as Julia Roberts and Scarlett Johansson. And why on earth aren't we following the lead of some innovative college courses and teaching our high school science students about science via the movies? Learning how science is often *wrong* in such big-screen depictions, and how scientists themselves do not fit Hollywood stereotypes, would be highly illuminating and memorable. See, for instance, the University of Central Florida's "Physics in Films" course, http://www.cah.ucf.edu/news/2004-Physics-in-Films.php.

125 *Merely straddling the line between physics and chemistry:* See Brian Vastag, "Assembly Work," *Nature*, Vol. 453 (May 2008), pp. 422–423.

125 *IGERT:* As we composed this book, 195 IGERT grants had been disbursed, averaging just over twenty per year. Yet each funding cycle, the National Science Foundation reports receiving between 400 and 500 preliminary IGERT proposals, only 20–25 percent of which even make it into the second stage because of current budgetary constraints. Congress should act now to dramatically expand funding for the program.

126 *pent-up scientific talent:* Addressing the plight of postdocs requires its own specific set of solutions. Many of these researchers in development express the concern that by embarking on the road less traveled, by way of congressional fellowships in policy, for instance—such as are offered by the American Association for the Advancement of Science's excellent program—they could harm their careers, because they need to publish as much research as possible to win an academic job. Instead, we need to foster the recognition that working in the legislative and executive branches of government inherently makes them better *scientists* in the broadest sense of the term, because there is no better way to grasp how science influences society. It would behoove the nation if such opportunities drew in more talented scientists, rather than deterring them. And just as a fellowship program currently exists to bring scientists into the federal government and onto Capitol Hill, similar fellowships should bring them into the media, the entertainment industry, and the religious community.

Conclusion

127 *the multibillion-dollar Large Hadron Collider:* Parts of the opening to this conclusion are based on a column by Chris Mooney for *Science Progress*, entitled "Cultural Collisions," September 24, 2008, http://www.science progress.org/2008/09/cultural-collisions/.

128   *"We've arranged a global civilization"*: Carl Sagan, *The Demon-Haunted World: Science as a Candle in the Dark* (New York: Ballantine Books, 1996).

128   *pills that significantly lengthen the human life span:* This is not necessarily as far off as you may think. See Robert N. Butler, Richard A. Miller, Daniel Perry, et al., "New Model of Health Promotion and Disease Prevention for the 21st Century," *British Medical Journal*, Vol. 337, July 19, 2008, pp. 149–150, noting: "Investigating how genetic mutations influence the basic rate of ageing is likely to provide important clues about how to develop drugs that do much the same thing."

129   *synthetic "telepathy":* Again, perhaps not as crazy as you may think. See http://cnslab.ss.uci.edu/muri/index.html, noting, "We aim to process EEG and MEG signals to determine what words a person is thinking and to whom or what location the message should be sent."

132   *"We require a common culture":* C. P. Snow, *Public Affairs* (New York: Charles Scribner's Sons, 1971), p. 10.

INDEX